# Learning to Lead Together

Never before has there been such strong recognition of the importance of community-based green spaces to local communities and urban redevelopment. This book is an autoethnographic account of the leadership dynamics behind setting up a charity and creating community gardens in unlikely places in the middle of London. The interwoven stories provide first-hand, examples of how an ecological and relational approach to organisational development and urban regeneration helped shift the business as usual paradigm. It will help you identify and step beyond individualistic and "heroic" notions of leadership, and will inspire you to find your own way of embracing natural and shared authority.

The book focuses on the experience of establishing and developing the work of environmental education charity Global Generation. It shows how action research, nature connection practice, and storytelling has grown shared purpose, trust, and collaboration, both within Global Generation and in the wider community.

The style and structure of the book reflects the participatory approach that it presents. The author, Jane Riddiford, deliberately challenges the separating norms of academic writing which is all too often authorless, dry and "othered". Going beyond this narrow framework the book combines different styles of writing, including traditional and autobiographical storytelling, diary entries and co-writing. Along with practice accounts of what happened, challenges raised and lessons learned, each chapter will also include other people's descriptions of their experience of being involved in the process.

**Jane Riddiford** co-founded Global Generation in 2004. She has more than 30 years of experience in setting up and running environmental, arts, and educational projects within urban communities in New Zealand and the UK. Her doctoral research focused on leadership framed within stories of ecology and cosmology.

# Learning to Lead Together

## An Ecological and Community Approach

*Jane Riddiford*

Routledge
Taylor & Francis Group

LONDON AND NEW YORK

First published 2021
by Routledge
2 Park Square, Milton Park, Abingdon, Oxon OX14 4RN

and by Routledge
52 Vanderbilt Avenue, New York, NY 10017

*Routledge is an imprint of the Taylor & Francis Group, an informa business*

*British Library Cataloguing-in-Publication Data*
A catalogue record for this book is available from the British Library

*Library of Congress Cataloging-in-Publication Data*
A catalog record has been requested for this book

ISBN: 978-0-367-68432-7 (hbk)
ISBN: 978-0-367-68433-4 (pbk)
ISBN: 978-1-003-13751-1 (ebk)

Typeset in Bembo Std
by KnowledgeWorks Global Ltd.

Printed in the United Kingdom
by Henry Ling Limited

To the children and young people who
are the life blood of our work

# CONTENTS

# ACKNOWLEDGEMENTS

"Weaving in and out, down the lines and round again." This was the fiddled song of the ceilidh caller on our wedding day. Not only was it a celebration of the weaving together of my husband Rod and I, as we were both a few years out from being young, it was very much a celebration of people from different parts of our lives. I feel the same thing now; the creation of this book has been a weaving together of hemispheres, eras, ways of being and most of all of people to whom I owe many thanks.

When I was interviewed for ADOC, the Ashridge Doctorate in Organisational Change, I rather apologetically said that for me it would need to be a collaborative endeavour, I couldn't think of anything significant I had done on my own. I was told, that's what was expected. At the time I had no idea how many different people would be involved and how much I would need their support. Much of the material and certainly the approach that gave birth to this book comes from that endeavour. Many thanks go to my supervisor, Dr Steve Marshall, for giving me the space to find artistry and an interest in the tricky terrain of leadership and for his perfectly timed challenge and support. Steve along with the rest of the faculty, Drs Kathleen King, Gill Coleman, Chris Seeley, and Robin Ladkin instilled in me an appreciation for the transformational power of action research. I went on to supervise MSc students with Gill Coleman on a Middlesex MSc in Transdisciplinary Studies. Working with Gill and supporting a cohort of inspiring community change makers in their own action research journey has been a privilege and developed my appreciation of participatory ways of working and personal writing that can bridge every day and scholarly settings. Some of the stories that found their way into the following pages were first crafted as stories to tell and for that I am grateful to have spent time with storytelling teacher Roi Gal-Or at the International School of Storytelling at Emmerson College in Sussex. Roi emboldened me to attempt weaving autobiographical stories into traditional myths and wonder tales.

The dedicated team of staff and volunteers at Global Generation are the backbone of this book, they are what made the inquiry and the community projects behind it possible. Together we have worked with many young people, made gardens in unlikely places and we have dived into an on-going exploration of what an organisation inspired by the rhythms and patterns of nature might actually be. Special thanks go to Nicole van den Eijnde, Silvia Pedretti, Rod Sugden, and Rachel Solomon for helping me establish action research as a way of being in our work together. Thanks also to Emma Trueman, for the painting on the cover and the 14 Billion year journey illustration. Ballast and grit behind our work has come from the often invisible trustee board. Big thanks to Paul Aiken and Sarah Riley who took a chance and joined me as founding trustees and company directors when we first set up our organisation in 2004 and to Lela Kogbara, Tony Buckland, Jane Jones, and Steve Marshall (who after being my academic supervisor became our chair of Trustees) for engaging in the final chapter of the book, which was the hardest one to write.

The New Zealand back story that weaves in and out of these pages came through family archives and conversations. It has been a privilege to collaborate with three generations of my family; my mother Yvonne Riddiford, my sisters Liz Riddiford and Lucy Riddiford and my nephew Sean Conway; their interest gave me the courage to dig deeper.

Invaluable support came from Professors Mānuka Hēnare and Chellie Spiller who were then at the University of Auckland Business School. They helped me identify my Tūrangawaewae; the ground I stand on and through their guidance I felt able to weave the Māori legend of the Three Baskets of Knowledge into my story. Chellie ran a Wayfinding Leadership workshop for the Global Generation team and our associates in 2016 confirming and helping us develop an approach which felt natural and stood us in good stead for the challenges that lay ahead of us. Mary Evelyn Tucker and John Grim, professors at the Yale University School of Forestry and Environmental Studies and Divinity School respectively, have backed and sustained an ongoing interest in the way colleagues and I have brought stories of the universe, largely inspired by Thomas Berry and Brian Swimme, to children and young people in the middle of London. A link in the chain was my discovery that they had collaborated with Mānuka and knew Chellie, this gave me the confidence to find my way into what I sometimes call cosmic indigeneity; a universal identity available to all. A significant step in this story of connection, was the introduction made by Mary Evelyn and John to Jonathan Halliwell, Professor in Theoretical Physics at Imperial College London. Jonathan read early drafts of many of these chapters and encouraged me to keep going in my loose and somewhat creative interpretations of the scientific origin story as a meaningful creation story and a useful narrative for community-building.

Initial editorial support came come from Jocelyn Vick Maeer, who I first met as a 13-year-old on a garden project Global Generation was helping with on a housing estate in King's Cross. Many years later, Jocelyn worked for Global Generation, by then she was an anthropology Masters Graduate and a committed social and environmental activist. I soon discovered that she was also a very good writer. Along with editorial suggestions, Jocelyn gave me hope that the book might be of interest to young activists. Creative non-fiction author, New Zealander, Pip Desmond whose books I have devoured, also read and made suggestions about the manuscript. Thanks to the Routledge editorial Team; Rebecca March, Sophie Peoples and Roshni Bhandari for making this such an enjoyable endeavour.

Many others are named and have added words to the pages that follow to whom I owe thanks; architect Jan Kattein's creative commonplan process and lawyer Ben Jones's gentle crafting of legal agreements provided a vital leg up in the early stages of both the Skip and Story Gardens. Property developers Argent, British Land and Stanhope have provided land and some of the financial resource for the gardens and some of the activities that happened in them. Special thanks go to Roger Madelin and Juliette Morgan who have read and commented on chapters and most importantly they both believed in and backed our unconventional way of doing things. Local Somers Town resident TJ (Tania Jacobs), brought valuable nuance into my understanding of the often fraught relationship with construction companies in places (like most parts of London) where large scale regeneration is occurring.

Finally, I am deeply grateful for the patience and creative input of my husband and co-conspirator Rod Sugden who is my rock and has been an ongoing sounding board for the book; he has helped me stay true to the spirit behind our work together.

# THE STORY OF HOW THIS BOOK WAS WRITTEN

I remember picking blackberries beside Wirihana, the Girl Guide camp house next door to where we lived; I must have been about ten. We used to put the ladder on top of the brambles and scramble over to find the sweetest, juiciest berries. I loved the wildness of the place. One day I was there on my own and, when I had picked enough berries, I lay on the grass. Feeling the warmth of the sun, I looked up at the sky and the space. I was pulled by a sense that there was more going on than I had been told about. What was it about the space that went on forever?

Years later I came to understand that space as the endless field of emptiness out of which everything, all the objects that I have names for, arises. The sense of a deeper reality that flows through all things made me curious about who I am, beyond the more obvious labels I gave myself. At an early age, I glimpsed that I wasn't an isolated individual and as a young adult this creeping sense that I was part of a bigger stream of life sometimes brought fear as it clashed with the commonly accepted arrangement of what was important in life; get a job, get married, have children, grow old, die. Perhaps I felt this more acutely in a secular place like the New Zealand of my growing up years. A land of contrasts; big hearted and self-deprecating, easy going and politically active and the place in which I was asked, more than once, if I could be just a little more superficial.[i] This tension was the catalyst for me to deviate from the script of what was expected. I ventured abroad in search of answers about connection and wholeness; a search that eventually brought me home. Many years later, this search inspired what I brought to Global Generation, the London based charity that I co-founded in 2004. I was drawn to what we can learn from the interconnected and evolutionary movements of nature and the fresh and imaginative way children and young people respond to nature if given half a chance. For that reason, I sometimes refer to myself as a social ecologist. The human dynamics involved in leadership and community-based regeneration are what I grappled with in setting up and running

Global Generation's projects and became increasingly curious about during the course of my doctoral studies. For many years, I was silent about the very cosmic story behind my enthusiasm and my contributions in the world. These experiences and the way they played themselves out in Global Generation are what this book is about. Three interwoven story-lines run through the following chapters:

- Leadership: The story of a woman who has grown towards a more collaborative style of leadership in a community-oriented, environmental education charity.
- Colonialism: The story of how the shadow of the past affects the present.
- Ecology and Cosmology: The story of how finding home in the earth and the wider universe can positively shape the underpinning narrative of leadership.

## Insider research

Through a late-in-life academic journey, I came across the world of action research, which is the inquiry process out of which the content of these pages came. Contrary to many forms of research where the author's voice is kept on the side-lines, action research is transparently personal. As Judi Marshall[ii] unashamedly points out, objectivity is not an option: instead, one seeks to develop a critical subjectivity.

The following description provided by Peter Reason and Hilary Bradbury in an early edition of The Handbook of Action Research has been referenced multiple times and, in my view, stood the test of time. It speaks to why this project, which I began more than ten years ago, has not left me alone.

> ... action research is a participatory, democratic process concerned with developing practical knowing in the pursuit of worthwhile human purposes, grounded in a participatory worldview which we believe is emerging at this historical moment. It seeks to bring together action and reflection, theory, and practice, in participation with others, in the pursuit of practical solutions to issues of pressing concern to people, and more generally the flourishing of individual persons and their communities.[iii]

The three territories of action research most commonly referred to are first person, second person, and third person, which can be simply thought of as; "me, us, and them."[iv] In the way I have written, there is a crossover between first person action research and what is known as autoethnography, a relatively recent anthropological discipline. Rather than studying other people, an autoethnographer uses the study of themselves as a way of gaining insight into culture. Additionally, storytelling is used "to illuminate larger social issues where the micro and the macro come together and illuminate one another."[v] Identifying with the ancient and emerging story of nature and the wider cosmos provided me with a pathway that married the more singular, human-centric approach of autoethnography with a broader participatory process of action research.

Alongside critical turning points in my work with Global Generation, I have paid attention to tiny moments of connection in the participatory life of my senses[vi]; the changing of the light, the sensation of wet grass under my bare feet, the feeling in my bones. Over the years, I simultaneously wore a number of different hats; organisational leader, storyteller, educator, doctoral researcher, and wonderer. This meant I didn't determine in advance at what point someone or something would become an object of research.[vii] Researching on the inside can be a messy process and runs the risk of betraying trust. I have been challenged about the validity of the writing; was it just from my memory? Sometimes it was and in that sense its value is subjective. Other material comes from recorded conversations, excerpts of emails, published texts or what people have written themselves. One colleague pointed out that at the beginning of my research they felt like a pawn, thankfully by the end they said they felt like a participant. Another made it plain that they wouldn't have spoken so vulnerably if they thought it would be written about; I respected his wish and left out that writing. Often, I worried that telling people I was engaged in research would create a barrier and suck the life out of a conversation. At the same time, I was aware of the need for informed consent. Inviting those I had written about to comment and to contribute to the writing themselves kept me honest. It was a way of creating a shared inquiry and in many instances gave voice to different perspectives. I am grateful for the generosity and open-mindedness of colleagues who became used to me pulling out a recorder at random moments or suggesting they journal and blog about our work together; much of their writing appears on these pages. (Real names have been used with permission.)

## Understanding the past

Reason and Bradbury write of the flourishing of individual persons and their communities. Embedded within this aspiration to create the conditions for positive action in the world is the fact that we are relational beings. In other words, we are not vacuum-packed monoliths that stand in isolation to everyone and everything else, including the forces from the past that have shaped us[viii]. Our actions are influenced by the families and cultural landscape that raised us. This called for an excavation into my heritage. I wanted to understand the sometimes-conflicting forces that moved within me; the opportunistic and single-minded drive of the pioneer and the pull towards deeper values of connection and wholeness. My mother, Yvonne Riddiford, now 95, is the Kuia (elder) of our family. She has long been a supporter of Global Generation's work and my research journey. Inevitably she became part of this inquiry. She carries the stories of past generations and is actively involved in the lives of her children and grandchildren. In doing so, she sometimes finds herself empathising with and occupying conflicting worldviews.

Seen through an ecological lens, the actions of some of my forebears in relationship to the land do not sit comfortably. Surfacing these stories is not to blame but rather to understand the conditions and motivations of the past in order to open up different choices about the future.[ix] At first, I wondered how best to share my writing with my mother. In a phone conversation, I told her that I felt anxious about how she would receive my version of aspects of our family mythology. After reading my accounts, she wrote to me and said, "You are telling it how it is. If you didn't feel vulnerable, your work would not be good."

## How the book is organised

This deep dive into the ways in which the shadows of the past are sedimented in the present informed how I thought about writing this book. In my mind's eye, I saw it as a tree, starting with the roots behind the good and the not-so-good things that have influenced who I am and my contributions in the world. Next came the trunk and mainstay: creating shared philosophical underpinnings, developing coherent educational practices and growing a culture of collaborative leadership with my colleagues. The last three chapters branch outwards and describe practical steps and associated internal and external tensions in setting up new community gardens in different areas of London. The first chapters are arranged thematically and the later chapters are organised in the sequential order in which events occurred. The chapters are written so they can stand alone.

In his small and potent publication, The Myth Gap, Alex Evans[x] points out that the environmental movement has relied on facts and figures, at the expense of the power of stories to change hearts and minds. I hope that activists and scholars alike will find this book useful, and to that end, each chapter includes a reflections section which explains some of the lessons I have learned. I also hope the stories, mostly autobiographical and occasionally mythical, will offer anyone an enjoyable read, no matter what you are involved with. Over the last 16 years, there are a multitude of stories that I am sure different people involved with Global Generation might feel it is important to tell. Whilst I have included the voices of others, this is my version of events and in that sense, it is not the whole story. It would be a good outcome if you as a reader were inspired to start surfacing and telling your own stories. By way of encouraging that I have included an appendix with invitations for your own inquiry that relate to each chapter. All of the prompts come from workshops my colleagues and I have led over the years.

## Community-based regeneration

The very localised stories I tell are microcosms of a larger cultural story. As I see it, the way we have treated land indicates the way we have suppressed and lost appreciation for a more spontaneous underlying reality. This is a reality that reaches beneath cultural lines, a reality that perhaps should remain un-named. I believe this deeper, untamed ground is a wellspring of creativity and connection. In the pages that follow I describe a childhood coloured by a fault-line; scenes of burnt hillsides, cleared of forest in the interests of farming. It is the back story to why I became interested in the effect of the modernist worldview on land, community, and leadership. I found my way spiritually and literally into a professional practice which I sometimes describe as drawing life out from beneath the concrete, in both place and people. In the early 1990s, I was involved in the creation of an Inner-City Forest which brought together the local community, on the edge of a busy motorway in Auckland, New Zealand. Now decades later, I am involved in similar community-based regeneration work in the middle of London, where I have lived for the last 25 years.

In May 2018, I began supporting a small cohort of mature Master's students who were on an action-research-based MSc with Middlesex University, with the focus on community-based regeneration.[xi] Participants included teachers, gardeners, storytellers, a vicar, a playworker, a local authority leader, and a town council manager. This was the first time I had supervised an academic programme. It felt like a huge responsibility and a privilege to work with the group who brought with them a wealth of experience, culture and professional roles which they inquired into and wrote about. Their writing took me into the rich terrain of the worlds they inhabited as they described experiences of community connection, the revelation of spirit and soul, black mental health, grounding in, and listening to the voices of nature. As they wrote and shared their papers over the course of two years, I worked in tandem, writing and sharing chapters of this book with them. The connection between us was a constant reminder for me to do what my colleagues Gill Coleman, Dave Adams, and I were asking our students do. In the spirit of scholarship, we wanted them to locate their personal experiences within the wider landscape of other people's ideas. However, most importantly we wanted each of them find their own voice by writing from the closeness of their lived experience.

On our last workshop, we discussed what we now thought community-based regeneration might actually be. Kate, who is the Community Development Manager for Frome Town Council, highlighted why regeneration is such a problematic word.

> The word brings connotations of traditional community development of 1950s slum clearance. This was very much about imposition and more about the physical environment than anything else. There was an attitude of we are going to sort out these communities in a very top-down way. It is the word that attracted me to the course and now I struggle with the word. In my own way, I too envisaged I could go in and help communities and move on and now I have moved away from a top down, philanthropic approach.
>
> Kate Hellard, February 2020.

The flipping of the monopoly board that is often seen as positive regeneration is rife in our cities; poor people cleansed from certain areas which are then developed to become desirable and expensive. At the same time, I like the word regeneration. It has currency with many of the organisations I work with and it is also a word that speaks to the restorative powers of the earth.

In the early 1990s, whilst studying horticulture, I learnt about the ecological story of gorse in New Zealand. When the European settlers arrived, they cleared the forests in their efforts to feed themselves, recreate the familiar and pursue global trade. They brought seeds of grass for sheep and cattle to graze and yellow flowering gorse, for hedges. In the warm southern hemisphere, gorse spread like wildfire. As a child, my memory is of long car journeys looking up at hillsides covered in the bright yellow flowers of the shrubby, prickly, hardy gorse. It was known across the country as a noxious weed. However, in time, gorse began to play a regenerative role. As a legume, the nodules on the roots of the gorse captured atmospheric nitrogen and the soil became nutritious. In open ground, gorse also provided shelter and whilst by no means a replacement for the loss of native mānuka, gorse served a similar function to mānuka as the nurse crop for the climax layer of the forest which would eventually succeed the gorse. Along with planting by thousands of volunteers across the country, seeds of the forest giants that had laid dormant in the soil for many years began to germinate. When I return to New Zealand, I am always struck by the hills around Wellington, once yellow, now clothed in the dark green mantle of native forest. Nature, given a chance, is bringing life back to the land. The story of gorse has provided an encouraging metaphor for much of my working life.[xii]

In experiencing the ways in which a healthy forest is full of diversity and mutually supportive relationships, I began to see that growing a forest is growing a community. Poet Gary Snyder puts it like this:

> When an ecosystem is fully functioning, all the members are present at the assembly. To speak of wilderness is to speak of wholeness. Human beings came out of that wholeness, and to consider the possibility of reactivating membership in the Assembly of All Beings is in no way regressive.[xiii]

In the discussion with the MSc group which I describe above, Lela, one of Global Generation's trustees, who was born in Nigeria and was formerly deputy chief executive of London Borough of Islington and is now one of three founding directors of Black Thrive Global, shared her perspective on why diversity is important for growing community:

> In any locality, communities have very different things to offer, they have different riches and so it is not a single story … you can never build a community if you exclude people of privilege. The same thing relates to race, a lot of people want a black-only community, and that might be a necessary thing but ultimately it is not building community or regenerating community – it is not a long-term thing. Because everyone has to change.
>
> Lela Kogbara, February 2020.

Global Generation's community gardens are distinct from the straight-lined precision of the surrounding glass and steel buildings. However, their magic is not only in their juxtaposition but in their connection with the many different people that live and work around them. They are places where everyone can be involved; be they a local teenager, a developer, a Bangladeshi mother, or a construction worker; they help to make the life that is beneath the concrete a little more visible. Global Generation developed through engaging with a diverse range of stakeholders. The invocation of an older, less controlled rhythm has not only been between my colleagues and I and the young people we work with, but also with the businesses with whom we collaborate. Our work has never been "us over them," community group over the developers, or vice-versa. We have all shaped and are being shaped by revealing a more coherent story that has grown between us. It is a hyper-local story which draws upon and adds to a larger weave. In this way, none of our gardens could have happened through a master-planning process. We often refer to these community spaces as gardens of a thousand hands, hands which crafted an unconventional assemblage of structures, which are in stark contrast to some of the corporate locations we operate within. They are structures that have needed to meet the requirements of planning permission and building regulations. Often as I spoke with developers, lawyers, architects, and educational establishments to draw up complex legal agreements, I witnessed belief in a lesser-known and more enchanted world that lies beneath the mechanical world. Consequently, agreements have been cleverly couched in phrases such as "with reasonable endeavour." I call that being multi-lingual.

## Leadership

Compared to all the espoused theories and numbers of people who are seen to be leaders, early on, I became curious about the fact that there are not many accounts by people writing about their own lived experience of leading. Recognising this gap encouraged me to keep noticing and keep writing about my experience. I hope the documentation of this journey will offer learning for leadership within community and sustainability projects and take out even in more corporate endeavours about the richness of uncertainty, the creativity of collaboration and most of all the importance of learning from the rhythms and patterns of nature.

Though some of the writing in the book goes back to 2008, I completed the manuscript in 2020 and by that time we were a very different organisation. From a band of two, we had grown into a team of 15 paid staff along with associates and volunteers. Leadership was not something I would have voluntarily signed up to nor was it a word I comfortably used. By sharing leading with others in the team, I discovered that leadership needn't be the domain of figures who command and control; most of all it was an ongoing inquiry. Stories drawn from ecology and cosmology provided a metaphorical foundation which helped me stand with curiosity in times of uncertainty and helped me embrace a process of leadership that is fluid and changing, sometimes singular and more often collaborative, particularly in recent years when I have shared the role of Global Generation director. This is how Nicole described our work together.

> My natural inclination is to shy away from the term organisational culture. It feels overused, a way to dictate and create sameness of opinion, a set of values for all to follow. However, at the same time, when asked what is of most importance to the running of Global Generation, I undoubtedly say that it is the people, how we work together, how we learn from each other, how we build trust first amongst each other and then how we expand this sense of community beyond into our wider community.
>
> Nicole van den Eijnde, GG Joint Director, 2018.

Not only were we a very different organisation, the world was a very different place. In 2006's Al Gore's film, An Inconvenient Truth, was not long out[xiv]. By 2020, in most circles (unfortunately not all) Climate Change was no longer a contestable issue. Then suddenly the world ground to a halt, in the grip of the COVID-19 coronavirus. Two thirds of the population were in lockdown, with an enforced call to stay home to save lives. With fewer cars on the road, London became quiet, the air noticeably cleared and more birds came in to the city. At the same time, huge numbers of people were dying and there was no knowing when the tide would turn. As I read through the manuscript, I asked myself how many of the ideas I had written about would stand up in the face of this global turning point. Over a matter of days, I went from panic to pause as I let in the seriousness of what was occurring. This was a call to stretch beyond a very real need to keep our organisation financially afloat into actively engaging with what would support the national effort to curb the virus. People all over the world reached out to each other; online, out of windows as we clapped for front line medical workers or with our neighbours in mutual aid and safely distanced street greening groups. Perhaps we listened more deeply in the knowledge that we were now walking in unknown territory. In our socially isolated safety bubbles, I noticed a greater sensitivity to the kind of intelligence that can only come through a collective. I prayed that this glimpsed potential might help a more connected, socially distributed worldview take root; one that is kind to both people and planet.

## Notes

i Chapman, M. (2020). *Jacinda Ardern: A New Kind of Leader.* Carlton Vic: Nero. Chapman describes how the nonchalant New Zealand psyche frowns upon the two qualities of enthusiasm and earnestness.

ii Marshall, J. (2004). *Matching Form to Content in Educating for Sustainability: The Masters (MSc) in Responsibility and Business Practice*, in Galea, C. (ed) (2004) *Teaching Business Sustainability: Volume 1*, Greenleaf Publishing, pp 196-208. Marshall describes the development of the MSc in Responsibility and Business Practice, of which I am an alumnus. The programme had "an explicit intent to 'address the challenges currently facing those managers who seek to integrate successful business practice with a concern for social, environmental and ethical issues."

iii Reason, P. and Bradbury, H. (2006). *Introduction: Inquiry and Participation in Search of a World Worthy of Human Aspiration*. In: P. Reason and H. Bradbury, eds. *Handbook of Action Research*. London: Sage, pp. 1-14.

iv Coleman, G. (2011). *Sustainability as a Learning Challenge, 360°* The Ashridge Journal, Summer, pp. 8-13

Marshall, J. and Mead, G. (2005). *Editorial: Self-Reflective Practice and First-Person Action Research*. Action Research, 3(3), pp. 235-244.

v Smith Sullivan, K. (2008). *The autoethnographic call: Current considerations and possible futures*. Ph. D Thesis, University of South Florida. Available at: http://scholarcommons.usf.edu/etd/503/ [Accessed: 26 June 2015]. An energising autoethnography conveying the important role of good storytelling and beautifully crafted writing. Despite being a study of well-known second generation autoethnographers like Caroline Ellis, Smith-Sullivan manages to evocatively tell her own story.

vi Abram, D. (1997). *The Spell of the Sensuous*. New York: First Vintage. In the spirit of phenomenology in this this ecological work the reader is invited into full bodied experiences of the natural world

vii Coghlan, D. and Shani, A.B. (2005). *Roles, Politics, and Ethics in Action Research Design. Systemic Practice and Action Research*, 18(6), pp. 533-546. A description Insider Research and the predicament I often find myself in, "If action research is a 'journey' and 'evolves' how can informed consent be meaningful? Neither action researcher nor participants can know in advance where the journey will take them and cannot know to what they are consenting." P.340

viii Ladkin, D. and Spiller, C. (2013). Introduction. In D. Ladkin and C. Spiller, eds. *Authentic leadership: Clashes, convergences and coalescence*. Cheltenham: Edward Elgar. pp. 2-3. Donna Ladkin and Chellie Spiller, challenge the notion prevalent in much of the authentic leadership literature that promotes the notion of self as an autonomous inner essence, this they say "denies the patina of forces that interplay to forge a person".

ix Conquergood, D. (1985). *Performing as a moral act: Ethical dimensions of the ethnography of performance. Literature in Performance*, 5 (1), pp. 1-13. In researching my family history, I have tried to avoid falling into the ethical pitfall that Conquergood refers to as the 'custodians rip off' in which cultural custodians ransack their own or others' biographical past, often denigrating family members. P.5

x Evans, A. (2017). *The Myth Gap: What Happens When Evidence and Arguments Aren't Enough*. **Cornwall**: Eden Project Books

xi The Masters cohort with a Community Based Regeneration foucus, that I refer to were part of a wider MSc in Transdisciplinary Studies, run by Global Generation and Middlesex University.

xii Riddiford, J. (2015). Belonging in the Cosmos. In: C. Spiller and R. Wolfgramm, eds. *Indigenous Spiritualities at Work: Transforming the Spirit of Enterprise*. Charlotte, NC: Information Age Publishing, pp. 117-135. Chellie Spiller and her colleague Mānuka Hēnare helped me learn about and appreciate aspects of Māori cosmology, some of which spoke to my experience of evolutionary cosmology, particularly the works of Thomas Berry, Mary Evelyn Tucker and Brian Swimme which I describe in chapter 5.

xiii Snyder, G. (1999). *The Practice of the Wild*. Berkeley: Counterpoint.

xiv An Inconvenient Truth is a 2006 American documentary film directed by Davis Guggenheim about former United States Vice President Al Gore's campaign to educate people about global warming.

# 1

# AN INNER-CITY FOREST

Amid the complexity and immobility of the rocks, there rise suddenly toward me "gusts of being", sudden and brief fits of awareness of the laborious unification of things, and it is no longer myself thinking, but the Earth acting.

Teilhard de Chardin[i]

Donald Worster suggests that "mythologies create symbols and dreams for what to live by."[ii] In this sense, the stories in this chapter are personal mythologies. They contain formative events within my cultural background, which have shaped what I bring to community-based regeneration in multi-cultural contexts. Whilst I haven't always had the words for it, finding ways to address the long shadow of our separation from the land has been a motivation that has run through all of my working life. Māori are the Indigenous peoples of New Zealand and Pākehā refers to non-Māori New Zealanders, in my case, of English descent. I describe experiences which helped me as a Pākehā find a different relationship to the land than many of my New Zealand forebears had. I reflect on my experiences as a Pākehā from a pioneering farming family and my journey into aspects of Māori philosophy. This provides a personal lens for exploring how colonialism has affected me; a descendant of colonisers in my leadership role with Global Generation in the middle of London. To give you, the reader, an understanding of this lens, I share some background about my family and the form of colonialism and associated deforestation of land that occurred in New Zealand within the last 200 years. Writing in the style of creative non-fiction, with stories drawn from family history, has been a way of understanding mistakes of the past and also finding new possibilities for the future. In this sense, these stories offer possibilities for resolving inherited fault lines; whatever our cultural background might be.

## The ghosts of the forest giants

My childhood was coloured by the bush, the rivers, and the sea. A partly volcanic land, baked by a South Pacific sun, is a land of intense contrasts. Deep blue sky that goes on for miles framed by sharp silhouettes of rising and falling hills making shapes and telling stories in the distance. Stories that would take me many years to understand. One of my earliest memories is looking up at a heavy farm horse who towered above me and above him was a small bush-clad mountain called Titoke. His name was Roanie, such was the colour of him. I must have been about 4 or 5 years old. As often happened when we were with the horses, I waited until I felt my mother's arms around me, she picked me up and tucked me into the curve at the back of the saddle behind my sister, Liz. Just like we were in the photograph below, I held on tightly and we headed off. My other siblings and various cousins were doubled up on horses on either side of us; Starlight, Bill, and Bluey were their names. The smell of my sister's shirt pressed against my face, the sound of hooves on the gravel was both exciting and comforting. Next came the waters of the Waimahora stream splashing. We cantered up the steep bank on the other side and I looked out and saw the sheep and the black cattle grazing. Then I saw something that haunts me to this day. All around, dotted through the landscape, were the black and charred forest giants: rata, rimu, and totara. These were the ghostly remains of the native forest that had been burned in the interests of laying down pasture for farming. We made our way through the river flats and up the other side of the valley. Arriving at a rickety old fence of totara posts with barbed wire in between, we tied up the horses and scrambled over. From an early age, I learned to go off the path, climbing to the wilder lands beyond the fences[iii]. In the bush, it felt good and I liked to get lost there. The native grasses underfoot were soft and smelt sweet, fantails flew around our heads and the light trickled through the canopy of ferns.

Jane aged 4 and sister Liz on Roanie at Rangitoto farm, at the head of the Waimahora valley in the King Country, New Zealand 1967.

I always looked up to my sister, Liz; she knew all the names of the native trees. By the time I reached secondary school, she had gone on to university, but her name still hung in the memories of my teachers. I never thought I would be as bright as her, as pretty as her or as sassy as her. She was at law school, had a heavy truck drivers licence and drove a big red motorbike. School for me was pretty much about looking out the window. I did make it to university, but only lasted a year before I was called to that place beyond the fences. I dropped out, joined a theatre group that worked outdoors on stilts and then left New Zealand in the guise of being a groom for racehorses. Travelling to Australia, I went for free with Flying Tigers, a stock delivery airline.

This search took me to England. In time, the pull towards a deeper sense of meaning and connection led me into an encounter with a spiritual mentor which was to unfold and eventually unravel. It is a tale of wonder and promise. It is also a tale of misplaced heroics and abusive power dynamics that I shall return to later. Within that story and in my view, within many current stories about leadership gone awry, lies a deeper question posed by Australian eco-philosopher, Freya Mathews. How do we treat the ground beneath our feet? What is the attitude of modernity to the ground on which we walk and live?[iv] Mathews goes on to describe how modern civilisations have treated land as a commodity to own, fence, and farm. The communities of plants, mycelium, and bacteria that breathe within land are reduced to a neutral substrate on which to manufacture and impose our own designs. Sub-divisions into blocks of separate parts have become the dominant refrain of an increasingly bland and lifeless landscape. A controlling picture of modernity that also shows up in the regeneration of our cities.

## The return of the forest giants

I felt called to return to New Zealand; at the time, I didn't know the real reason why. Years later, I would frame this move as a return to the land, an opportunity to find some kind of resolve for the fractures I experienced within the land. I stayed with my sister, Liz, and on the first day, she gave me a brass key. "Don't lose it", she said. Each day, I would feel in my pocket and handle its weightiness. One day returning to the house early when no one was home, I felt for the key and it was gone. Liz's words returned to me, "Don't lose it." Panic filled my mind. I wanted to get away from the house and headed off on my small red scooter. Making my way along the edge of the motorway, I discovered a little clutch of houses nestled into the hillside. One of them took my attention. It was made of corrugated iron and outside was a for-sale sign. Over the coming weeks, I kept returning to look at the house and eventually plucked up courage to knock on the silver painted front door. It was opened by two smiling women who invited me in. Bridget and Sally had built the house, and planted a thriving garden full of native trees, including rata, rimu, and totara. That house became my home and it was a special place to be. The only problem was the noise of the motorway. Sometimes, I would close my eyes and try and pretend it was the sea. At other times, I would look across the Newton Gully where the motorway snaked through, and imagine what it would be like if the clay bank that had been stripped bare by the development of the motorway was once again covered by the mantle of dark green bush. Over the coming months, ideas turned into actions. The bank was part of the grounds of Newton Central School. My neighbour was a school governor and she introduced me to the head teacher, Tim Heath. He too had been dreaming about a return of the native forest to the denuded bank beside the motorway. Very quickly, Tim agreed for me and a friend Maurice Puckett to lead a significant re-vegetation project involving all of the children in the school and members of the local community.

Two Māori tribes, Ngāti Whātua and Tainui, had historical connection to the land this part of Auckland was built on. It was an honour and a good omen that they agreed to come together to bless the land as a ceremonial start to what became known as the Inner-City Forest[v]. From that day, it became an enchanted forest; a place where different people came together with each other and the natural world. Children were allowed out of their classrooms and through growing the forest, they learned to follow the rhythms and patterns of nature. They collected seeds from a nearby remnant old growth forest and grew them around the school swimming pool which became a community plant nursery. Together, we covered the bank in logs and branches and began a gentle and effective process of restoring the soil. We planted manuka as a nurse crop and over time, we created the conditions for the forest to grow.

One of the unexpected benefits of establishing the forest in this way was that a community of people connected to the forest began to grow. One of these was a retired farmer called Buckley Fyers. In his later life, he developed a passion for weaving kete; traditional baskets made of harakeke (flax). At the time, this was an unusual thing for a Pākehā man to do; Māori custom meant that the sacred art of weaving was usually done by the women. Buckley gained the trust of well-known weavers from all over Aotearoa, New Zealand, and in return he curated a collection of harakeke with a detailed description of what kind of weaving each variety was suitable for. He donated 24 varieties to the Inner-City Forest, and they are growing there to this day.

We developed a long term management plan for the Inner City Forest. This was created as a future resource for children of the school and their children's children. The image on the next page shows one of the ways our plans were brought to life by children's artwork.

Eventually I returned to live in London and there was a gap of ten years before I would visit Auckland again. When I returned, I was staying with my younger sister Lucy and I got up early to make a pilgrimage to the forest. I made my way through the knot of motorways which is an unfortunate feature of that city. Wondering what I would find, I felt a knot in my stomach. Then suddenly I heard the song of a tui and on a branch nearby a fantail flew. Ahead I could see, poking through the newly formed canopy, those forest giants: rata,

Cover of the Inner City Forest Management Plan.

rimu, and totara. Walking along the winding pathways; it felt good, it smelt good and I knew the forest was being cared for. The Inner-City Forest is now an important feature in the life of the school. It is also now protected by the New Zealand department of Conservation. I maintain that the opportunity it gave me to walk in the ways of the forest is the foundation for the community-building work I have done with Global Generation in the middle of London. As I mentioned previously, it is a practice that I often describe as drawing life out from beneath the concrete, of both land and people.

## Understanding the past

When speaking about the development of Global Generation's work, I often describe how you can't grow a forest without roots. This means that the most important part of our work is the least visible aspect, the part that lies beneath the soil. It is not the technical steps of designing workshops and growing vegetables, it is knowing ourselves and what we each bring to the work that determines how the work will flourish. The question of what we bring to our work opens bigger questions of identity; who am I and where do I come from. Philosopher, Daniel Taylor, describes how the past is "sedimented" in the present and elaborates that we are "doomed to misidentify ourselves, as long as we can't do justice to where we come from."[vi] In 2008 and 2009, Judi Marshall, Gill Coleman, and Peter Reason were my tutors on an action-research-based MSc in Responsibility and Business Practice at Bath University. In their book, Leadership for Sustainability, they describe how if we are to create sufficiently robust change in the world, we must question the ground we stand on. This, they say, will help us find the agency, awareness, resources, and approach to take appropriate action.[vii] This echoes the value New Zealand Māori place on knowing our past. Māori Scholar, Ranginui Walker, describes the importance for Māori of finding natural authority through knowing our Tūrangawaewae, which he explains is "the standing and identity of a people."[viii]

I have not always found it comfortable to be a Pākehā inquiring within this space. It sometimes feels like walking in land-mined territory. One must walk on one's own ground, which can feel like a very thin and fragmented line. It is not about looking out there, it is about looking at myself. The shadow of colonialism is long and insidious. It is not always easy to identify how conditioned in colonialist ways of being I actually am. Becoming a little bit aware can make it harder to even begin. One is bound to get it wrong, to upset people. Back in the early 90s, when we were planting the Inner-City Forest, I didn't have the words to describe the discomfort I sometimes felt. The awkwardness lay in the fact that at times being a Pākehā felt like an obstacle. Whilst growing the forest was the most natural thing to do and the place where I felt most at home, I felt that I couldn't presume a connection to land that went beyond the practical. It took many years for me to understand why and to feel confident in finding my place in the land.

Many years after the Inner-City Forest had become established, my sister, Liz, and I walked across the land where I had spent my teenage years. We were on our way back from an early morning swim in the Ruamahanga river. We spoke of Tūrangawaewae; the ground on which one feels empowered and connected. As we talked, we watched the sunrise outline the sharp upward-reaching shapes of the Tararua ranges in the distance. Our conversation went deeper when my sister said, "Rather than exploring the source of your inspiration, it's about actually seeing where you stand right now. Tūrangawaewae for Māori is the place where they know who they are. This includes your family grouping and the land. You, Jane, know where you come from, but where do you stand now?" I sensed this was pointing to ground from which change could grow. For me, this has meant looking more closely at the conflicting feelings of connection, needing to get away and grief that I have experienced in relation to New Zealand. As New Zealand historian, Michael King, describes in his book "Being Pakeha,"[ix] I needed to understand the soil from which I sprang and the past that explains, at least in part, "the emotions of the present." Much of the critique of colonialism has come from those who have been colonised. Colonialism has also affected people like myself who are the descendants of colonisers. Some of my ancestors arrived in New Zealand in 1840. In England, they did not own land. Possibly because of the harshness of the enclosures and the displacement effects of the industrial revolution, their hopes for a better life lay in the promise of New Zealand as a Garden of Eden,[x] where they, like others of similar circumstance, could become landowners. The early settlers arrived to find a people, New Zealand Māori, who were steeped in a mythological way of thinking, in which land and people were not separate. In using "mythological" here, I mean working with metaphorical stories that connect us to intangible forces and help us find our place and purpose in the cosmos.[xi]

Early European settlers to New Zealand brought with them notions of human supremacy over nature. Many of them were driven by the desire to "reshape the land to make it fit for participation in new forms of global trade."[xii] This was a recipe for owning and stripping the land bare in the service of farming, which has had generational consequences for land and people. This sentiment is reflected in the diary of my grandfather, Spencer Westmacott who is pictured in the photograph on next page, riding across his farm looking out on a burn he had carried out some 12 years previously. Prior to the outbreak of the First World War, Spencer worked with a team of men to "break in" the bush in The King Country, one of the last areas of New Zealand to be settled by Pākehā.

Spencer describes the experience of his first burn, where several hundred acres were felled to make way for pasture: "'Let her go!' I called, seeing they were ready and, lighting the leaves of a prostrate tawa, I prayed now that after all our anxiety and my bushmen's efforts, we would be rewarded with a nice clean burn."[xiii]

Spencer slept uneasily that night, fearful the burn would not take, which would mean losing the last of the family money that had backed his farming endeavours. He woke to find the air thick with smoke and joined one of the men in his party who turned to him and said: "The whole bloody country is afire!" Spencer's spirits rose, "The fire had caught alright. Immediately before us was a black burn. The ground covered with ash, the bigger logs and branches charred and smouldering."

My Grandfather Spencer Westmacott on Peter, circa 1936

My own childhood recollections of riding through the steep slopes of the land farmed by Spencer and then his son, my Uncle Pat, are reminiscent of geographer, Ken Cumberland's description, "Hillsides were littered with fire-blackened logs and gaunt, dead stumps of forest giants."[xiv] This description is echoed by my mother, Yvonne Riddiford pictured as a child in the painting on next page. Years later Yvonne wrote about her bush childhood:

> Behind the house the face of Rangitoto, ("blood red face") would glow red in the sunset. It, and the surrounding hillsides had been cleared and grassed, but the fern was struggling to take over, with wineberry, "mother of the bush" as the first tree to regenerate, in the gullies. Everywhere stood the tall ghostly half burnt skeletons of the forest giants; rata, rimu, and totara which had withstood the original burn. They were to remain a distinctive part of the landscape for many years to come.[xv]

My mother Yvonne and her sister Margaret circa 1930 in the bush at Rangitoto, painted by their father Spencer Westmacott

In looking at my background, I have tried to resist the binaries of good and bad because they tend to entrench one's position. However, it felt important to understand the feeling I had of standing on divided ground within myself. There were different drivers at work within me; the opportunist spirit of the pioneer and a call to listen more deeply to the land. This division caused confusion which led me to leave New Zealand. However, it was in returning to New Zealand that I found understanding and some resolve with the fault lines of the past.

That morning after our swim in the river, listening to my sister, a glimpse of an underlying fault line opened inside me. I said, "Living away for the last 20 years, I have maintained a romantic relationship to the can-do, pioneering spirit that is so prevalent in New Zealand. However, within the long shadow of opportunism is the reality of land that has been stripped bare. This poses a clash with deeper currents of connection that run within me, "Sometimes, it feels like there are two parallel drives running in me. I don't feel fully connected as there is this division going on." My sister took the conversation further and said, "It's like an earthquake, the ground you stand on is shifting, deeper forces are at play. Up until now, you have had a nice little story about where you come from, before you talk about who you are now. Whereas the sort of work you are doing with young people and the things you are grappling with in building community need to be reconciled with motivations that lie at a very deep part of your being."

The truth of what was being said landed. "Yes," I replied, "I need to find out where the rifts that are within me come from; what are the deeper seismic shifts?" My sister agreed and said in a quiet voice, "It is something that Pākehā struggle to do … we are talking about fault lines."

## Honouring our roots

In 2017, Chellie Spiller, a Māori Scholar and co-author of a book called Wayfinding Leadership,[xvi] who I had corresponded with for a number of years, and her Mother Monica ran a workshop for the Global Generation community in the King's Cross Skip Garden in London. We learnt about practices that were handed down from early pacific explorers, who braved treacherous waters to unknown destinations. Deeply attuned to subtle nuances and changes within themselves, their team, the skies and the sea, they felt into the connection between things; continually resetting their sails and finding their way forward without maps. This way of being fully present in the unknown and travelling both hopefully and purposefully runs counter to a tight project planned process advocated within many leadership approaches; however, it was inspiring and deeply confirming for many of us who attended Chellie's workshop. The workshop opened with a traditional Māori welcome which is known as the karakia. Hearing this time-honoured invocation in the middle of the tall glass and steel buildings in this busy part of the City brought me home to a deeper part of myself. I felt all of my senses expand and the different threads of my life coming together; Aoteoroa New Zealand, my work in London and the hard to name but very present source behind my work.

In keeping with Māori custom, Chellie invited us to begin the workshop by taking part in the Māori practice of mihimihi, an age-old tradition of remembering who you are by naming one's connection with others and particular places in the natural world that have shaped our lives. Chellie describes it like this: "The mihimihi reminds humans that we are a movement through time, each a present link to the past and the future, and each woven into the fabric of belonging. Sharing one's genealogy, including one's name, ancestral ties, and connections to place, provides a platform for connecting and building reciprocal relationships."[xvii] The Māori word for this all-encompassing sense of genealogy is Whakapapa. It begins with Te Kore, the empty place of pure potential before anything ever happened. I hear Whakapapa as an invitation to experience connection with the whole of life. To know about a tree, a rock, or the beginning of time is to know my whakapapa.

The workshop in the Skip Garden attracted 25 colleagues and other friends of Global Generation who came from a variety of cultural backgrounds and cultural identifications, including Irish, Nigerian, Italian, Ethiopian, British, Eritrean, Scottish, Australian, Taiwanese, Dutch, and Pākehā. This group of people included community gardeners and chefs, academics, a quantum physicist, an actress/beekeeper, a vicar, a filmmaker/photographer, and social entrepreneurs.

Many of the participants were involved in some form of social or environmental activism and most were familiar with notions of an interconnected and non-separate sense of self. They were also familiar with the fecundity of the groundless ground of emptiness as a dimension of potential, which has strong resonance with Te Kore. For some participants, their sense of self included their evolutionary genealogy, as members of the Deep Time Journey network, extending back 13.8 billion years to the beginning of time. As Chellie noted, it was particularly powerful that we were with a group whose roots spread all over the planet in such an unlikely setting; a community garden in the midst of very corporate high-rise buildings. The workshop took us all back to our roots and had ripples that would guide Global Generation in the months and years to come.

I have always been surprised returning from New Zealand at how I also feel connected to England. I feel the connection as I walk the sacred ridge that flows up through Hampstead Heath, through the ancient Queens and Highgate Woods towards the heights of Muswell Hill; the site of several monasteries where no tube line was ever built. There is a stirring inside, even in writing this. My legs tingle as an older more original rhythm starts to come through. I have been here; I know this place. Following the Wayfinding workshop, the Global Generation team had four subsequent sessions with each other. Over many years we have regularly run enquiry sessions, which at the time we called 'Engaged Philosophy'. In the series of Wayfinding sessions, we focussed on the themes Chellie had introduced us to. In the first session, we began with the Mihi. This time, we had an opportunity to prepare over several days. In coming together, each of us recounted our lineage, starting with our ancestors and the land our families came from, through to our special places in nature where we now choose to spend time. For some, this meant going right back to the stars, "They are my guardians," said Rod. It was a privilege to sit in a circle with my colleagues in such a respectful and vulnerable way. We all felt we got to know more of each other and spoke about what it might mean for all of us to experience our indigeneity. Encouraged by the Wayfinding workshop, I have felt more interested and connected than ever to my British heritage and visited old Celtic places. Lela, who is from Nigeria, volunteered how much of an effect the workshop has had on her, in that it has made her realise that Wayfinding is her natural condition, and all these years, whilst working in leadership roles in local government, she has been made to feel that she should be leading in a more "certain, less changing and less responsive way."

## Reflections

Excavating and writing about the story of my past called me to learn more about myself. This included both the fault lines of my ancestors and a deeper more connected sense of self, which one might consider to be a more original way of locating self. Arguably, identifying self in the land and the wider cosmos is a way of closing the Western historical divide between self and nature. The early morning walk with my sister Liz to the Ruamahanga river began my quest for what the land had to say to me. Soon after, I wrote a paper called Fault lines about my experience in the land I was raised on. I felt some trepidation in sharing the paper with my family and also with Māori; I felt I might offend and upset people. However, it was through sharing my writing and meeting Chellie that I found resolve within inherited fault lines of the past. Writing this chapter and placing it at the start of the book, has been my way of bringing the spirit of the mihimihi into the following pages. It is a way of honouring my ancestors and more recent influences that are alive within me; including Liz, Chellie, and the land where I spent my childhood. Through the process of writing, I have grappled, and even come to empathise, with the actions of my forebears. Eco-philosopher Freya Matthews[xviii] maintains that without a sense of the reality of the past, we are not able to grasp our obligation to hand it on in any kind of a way that might be healthy or useful. "It is not by being virtuous but simply by passing their memories on to us that ancestors transform the neutral terrain of existence into a place of dwelling."

Receiving Buckley Fyers' gift of Harekeke was an invitation to learn about the practice and the significance of weaving, which has been a metaphor for my community regeneration practice ever since. I learnt that growing a forest is the weaving together of community. It was an iterative and a collective process. Not only did all of the plants need each other but the re-vegetation of the Inner-City Forest called upon the whole of the community – children and their parents, local volunteers, Kaumatua (elders) and neighbours all showed up to planting days bringing in a wealth of skill and enthusiasm. In many cultures, weaving is seen as a sacred thing to do, a way of bringing together the seen and the unseen dimensions of life to make containers that are both flexible and strong.

Writing personal stories in the way I have done can feel like a self-indulgent thing to do. However, if these stories offer a lens through which to understand a broader cultural landscape and open possibilities for change in that landscape, then they might be useful. This is when personal narrative becomes auto ethnography. After reading some of the stories contained in this chapter, my sister, Liz, wrote me an email which encouraged me to keep going with this way of writing.

> Weaving is the gathering together of threads and this is the role you play in the family. You are promoting a sense of connection with each of us. You are driven by the needle of your inquiry. You are taking it down and up, down and up through the warp and weft of the family fabric and this is opening up the inquiry for the next generation to break the mould and take risks.
>
> Liz Riddiford, 2013.

## Notes

i     Teilhard de Chardin, P. (1968). *Letters to Two Friends: 1926–1952.* New York, NY: New American Library, p.73

ii    Worster, D. (1977). *Nature's Economy: A History of Ecological Ideas. Second Edition (1994).* Cambridge: Cambridge University Press, p.20

iii     Akomolafe, B. (2017). *These Wilds Beyond Our Fences.* Berkely: North Atlantic Books

iv     Mathews, F. (2005). *Reinhabiting Reality: Towards a Recovery of Culture.* Albany: State University of New York Press, p.199. Mathews is an Australian philosopher working in the fields of ecological metaphysics and panpsychism

v      The Newton Central School website describes how the forest 'ngahere' is a designated 'Inner-City Forest' providing the school community with a bounty of taonga. It supports the reduction of the effects of car pollutants and regeneration of indigenous species of plant, bird, and insect life. The forest development commenced in 1993 with a whakatuwhera (blessing) and celebration by kaumātua from both Ngāti Whātua and Tainui iwi and by the local school community. Our forest is part of our school's everyday life, a place for learning, growing and connecting. Each child that has attended the school has planted a tree either individually, with friends, family and the wider community. New students bring a tree to plant and watch it grow as they do. It will stand as a tribute to their contribution to sustaining the earth and our future. https://www.newton.school.nz/our-forest/ [Accessed: 7 February 2020]

vi     Taylor, C. (2007). *A Secular Age.* Cambridge: Harvard University Press, p.29.

vii    Coleman and P. Reason, eds. (2011). *Leadership for Sustainability. An Action Research Approach.* Sheffield: Greenleaf. Sustainability, p.6.

viii   Walker, R. (1990). *Ka Whawhai Tonu Matou, Struggle without End.* Harmondsworth: Penguin, p.70.

ix     King, M. (1985). *On Being Pakeha: An Encounter with New Zealand and the Maori Renaissance.* Auckland: Hodder and Stoughton, p.125.

x      See Beattie, J. (2014). *Looking for Arcadia: European environmental perception in 1840–1860.* Australian & New Zealand Environmental History Network, 9(1), pp. 40–78.
        Fairburn, M. (1989). *The Ideal Society and its Enemies: The Foundations of Modern New Zealand Society 1850–1900.* Auckland: Auckland University Press.
        Park, G. (1995). *Nga Uruora: The Groves of Life – Ecology & History in a New Zealand Landscape.* Wellington: Victoria University Press.

xi     Doty, W. (2000). *Mythography: The Study of Myths and Rituals.* Tuscaloosa, AL: University of Alabama Press, p.15. Religious studies scholar, William Doty describes how mythic narratives are "not little but big stories, touching not just the everyday, but sacred or specially marked topics that concern much more than any immediate situation."

xii    Farrell, F. (2007). *Mr Allbone's Ferrets: An historical pastoral satirical scientifical romance, with mustelids.* Auckland: Random House, p.217. This historical novel, drew on a Turnball library receipt of the purchase of ferrets for importation to New Zealand, by my Great great grandfather, Edward Joshua Riddiford.

xiii   Westamacott, S. (1977). *The After-Breakfast Cigar: Selected Memoirs of a King Country Settler.* Westmacott, H.F. ed. Wellington: Reed, pp. 95-96. See "The After-Breakfast Cigar, my grandfather, Spencer Westmacott's memoirs, edited by his daughter in law my Aunt, Honor Westmacott, who married Spencer's son Pat Westmacott. Pat and Honor also lived at Rangitoto, the farm in the King Country of New Zealand where Spencer farmed

xiv    Cumberland, K.B. (1981). *Landmarks. Kenneth B. Cumberland tells how New Zealanders remade their landscape.* Sydney: Readers Digest, p.6. Ken Cumberland, Geography Professor at Canterbury University (1981, p.6).

xv     A Bush Childhood, an unpublished account of her childhood, written by my mother Yvonne Riddiford.

xvi    Spiller, C. Barclay-Kerr, H. Panoho, J. (2015). *Wayfinding Leadership: Ground Braking Wisdom for Developing Leaders.* Auckland: Huia

xvii   Quoted from a work in process paper co-written by Chellie Spiller, Fiona Kennedy and Jane Riddiford working title Thrown together: Belonging and recognition for a more sustainable world.

xviii  Mathews, F. (2005, p.179). Ibid/see above

# 2

# FROM ROOFTOPS TO DEVELOPER'S LAND

To re-inhabit the places in which we live is not to raze the smokestacks and freeways that we might find there, but to fit them back into the larger unfolding of land and cosmos.

Freya Mathews[i]

It might well be asked why someone in their right mind would lug tons of soil onto the top of an office building or into the middle of a construction site to grow a tiny patch of vegetables. I have never been under the illusion that Global Generation's efforts to grow food in London's inner-city areas would solve the issue of food poverty. Although the many hectares of bare rooftops and other empty spaces across London could massively extend the city's food growing capacity, the benefit of these projects goes beyond the vegetables. Growing food for people and pollinators in urban spaces is also a catalyst for opening up hearts and minds. Gardens, particularly those created out of reclaimed materials, provide many excuses for people to work together, to discover and express the best of themselves; arguably a good starting place from which to grow community. This chapter is hopefully a useful start for people interested in the practicalities of creating an urban community garden. I describe the places in which Global Generation's work developed over the first 12 years and the gardens that were created during that time. All of the accounts illustrate a way of developing projects that is an alternative to the tight, rational logic of a modern master planning process. It is a wayfinding approach that will probably resonate with many community projects, particularly if funds are in short supply. There is a kind of natural logic which has meant following inklings, sharing ideas with others, catching and fostering shared sparks of interest. The stories offer learning into how to find ways to begin community regeneration work and how to keep going despite, and perhaps because of, the inevitable bumps along the way.

This chapter also goes some way towards acknowledging the many people who have helped make the work flourish. For that reason, I have deliberately included their names and what they had to say about their experience of being involved in the projects I have written about. Of course, there are many more people who have helped, who I do not have space here to name; their contributions are not forgotten.

## More beginnings

The seeds for what was to become Global Generation were sown back in 2002, not long after I revisited the Inner-City Forest in Auckland. At the time, I was General Manager of a London-based arts organisation called Rise Phoenix. In the summer of that year, I was involved in taking a group of children on camp to Pertwood Organic Farm in Wiltshire. Many of them had never left the city before. In a big bus, we arrived at a small clearing in a gorse and Hawthorne copse, where we set up borrowed tents and a rather decrepit gazebo which was to be our camp kitchen. We were to be there for five days, and I remember that first evening, the children kept us awake, yelling, screaming, and talking all night. Many of them were terrified by anything that moved: the bugs in their tent, the owls, the rustling of the badgers in the undergrowth. I had serious doubts about what we were doing. Miraculously, after 24 hours, the atmosphere changed in the group and these city kids were soon absorbed by the forest, fascinated by the lizards and the spiders they found under the stones in the gorse. In the evenings, we sat by a blazing hot fire, humbled and safe in the canopy of stars. In the mornings, we walked across the dew-filled fields and found our place within a large stone circle. I felt as if the wind whispered in my ear saying, "Be still, the forest will not forget you."

The children's response to the stones, the stars, the fire, and the forest left a mark on me. Soon after the camp, I had pulled together a group of friends, including Sara Riley and Dick Crane who began to volunteer their time. At weekends, we would travel down from London to work on the campsite. One of the volunteers, Peter Rush, was an artist and taught at a nearby secondary school. For the first year, Peter lived at Pertwood in a yurt, bringing adventure and a certain elegance to the campsite and practically making it easier for us all. I would often pitch my tent beside an old yew tree: a tree I was to learn more about many years later. At the time, I was collaborating with Jane Goodall and our first big camp was to be a Roots and Shoots summit (the name of her youth movement). We brought together 20 adults and 50 children and young people, some from a country school and others from a private school who were involved with Jane. I brought children and young people from Kentish Town who were involved with Rise Phoenix and Odiri Ighamre and Paul Aiken (who was her husband at the time) brought children and young people from Kori Cultural Arts club, who were mainly of African and Caribbean descent. The youngest participant was 2 and the eldest, my mother, Yvonne Riddiford, was 80. Despite the many differences between us, we came together in a powerful celebration of each other and our place within the land. It gave me hope that a different, more connected world might be possible. It was this seed of hope I wanted to keep growing.

The volunteers and I were super excited; however, to my amazement, as I explained what we wanted to do with the camps, others in Rise Phoenix looked at me with confused and angry faces. Some of the trustees worried about taking children on camp and the artists didn't like all the new volunteers. Some of them insisted we were not an environmental organisation. I couldn't believe it. The more I explained, the more adamant they became. What could I do? In hindsight, what followed was a painful period of wading through dysfunctional organisational dynamics which I was poorly equipped to deal with. It was nonetheless a useful time of learning. To cut a long story short, the trustees of Rise Phoenix decided not to take responsibility for the campsite. Fortunately, the new volunteers were on board and were determined to keep the camps going. We wanted to bring the spirit of the forest and what had happened between a diversity of people to the middle of London. This was the catalyst for setting up Global Generation: the charity that would take the work forward. Sara Riley, Paul Aiken, and I became the three founding trustees.

We began to think together about the kind of organisation we wanted to establish. Whilst we were clear that we wanted to create an environmental education organisation, because of our experiences at Pertwood, our discussions focused largely on finding ways to break down barriers and unlock potentials within people. This is why we came up with the name Global Generation. We wanted a name that spoke of an inner change of perspective as well as the outer changes we hoped to achieve. The word global "indicates another perspective, a new consciousness, a different awareness altogether."[ii] It was this new level of awareness that went beyond boundaries, that we wanted to generate. All of us involved in defining the organisation felt that a core purpose was about growing a meaningful connection with the land; a connection that we felt could heal rifts between people. We agreed that this was not only about taking young people to the countryside but perhaps more importantly it was about unearthing the potentials of connecting to nature beneath the concrete in the middle of the city. We had all experienced the liberating power of encouraging young people to think about their roots; this included their deep time ecological roots in the land and their cultural roots. Odiri was from Nigeria and Paul from Jamaica, and in many ways, it was a blessing that they were involved with us. In looking back at that time, my mother said Paul and Odiri had a real ability to melt racism. One of the ways they did this was by encouraging those they worked with to stand fully in their cultural heritage, whether that was Somalian, Chinese, Irish, or English. In that way, we all had the sense of adding our own stories to a larger weave. The four photographs next were taken during camps at Pertwood. Not only are they of some of the key people involved in the early days of setting up Global Generation, for me they say something about the magic of what we were experiencing and why we felt compelled to set up an organisation to keep the work going.

Yvonne Riddiford and Jane Riddiford, Pertwood, 2004

Sara Riley, Global Generation co-founder, with Yoma and Amika of Kori Cultural Arts, Pertwood, 2004

Paul Aiken Global Generation co-founder and Dick Crane, Pertwood, 2005

Yvonne Riddiford and Paul Aiken drawing with children from Kori Cultural Arts at Pertwood in a session led by artist Peter Rush, 2006

## A living roof

Whilst digging up saplings for a tree-planting project at Pertwood, I started to imagine how we could bring the Pertwood experience to London. Our first real opportunity was, thanks to Charlie Green, founding director of The Office Group, who are providers of office spaces for small businesses in central London. Dick Crane introduced me to Charlie just after he had purchased an office building on Grays Inn Road in King's Cross. It had a flat roof and he had a sense that something special could happen there, although he didn't know what. In my mind's eye, I saw the potential of bringing the countryside to the rooftops of the city; they seemed like places where a pioneering spirit of the forest might grow. I hoped that the relatively unchartered territory of London's rooftops would encourage people to step out of their familiar worlds to come together to create something new. Charlie agreed to let us use his King's Cross office building as our guinea pig and provided some financial support, to which we added funding from grants. We discovered the concept of Living Roofs as championed by enthusiastic bird watcher Dusty Gedge. Dusty guided us through the early days of learning about membranes, load-bearing capacity, and recycled aggregates – including recycled sewage pellets and crushed glass from Canary Wharf.

From day one, people stepped forward to help us. A structural engineer gave us reports we could understand and actually implement. An ex-policeman and community safety officer for construction company Costain Laing O'Rourke organised cranes and road closures and in the nicest possible way helped us avoid some potentially nightmarish health and safety challenges. Moy Cash, the biodiversity officer for Camden Council wrote letters and spoke enthusiastically about the habitats we were providing for London's birds and beetles. The greatest surprise to all of us was when we did actually find an endangered ancient woodland species on the roof; a stage beetle *Dorcus parallelipipedus*.

This unusual finding occurred; thanks to the generosity of the Hampstead Heath Conservation team who brought us van-loads of sawn-up logs from the Heath, out of which we created wooden paving and habitats for insects. For a number of years, until the building was sold, the living roof was an important educational space for Global Generation's educational activities, as can be seen in the photograph below.

Zak Nur and other GG generators on the Office Group Living Roof – 2006 – Photographer Jane Riddiford

## Growing an organisational mythology

In the early days, working for Global Generation was unpaid or done for next to nothing. People helped because they believed that young people should be given a chance to make something significant happen. From the get-go, we involved young people in the process, wanting to give them the hands-on sense that they could create the future. Our voluntary workforce included ex-offenders from the Westway Project, as well as young people from Kori Cultural Arts Club, Camden Job Train, and our growing body of Generators (Youth Leaders) from schools in Camden. We learnt that what made a living roof was more than the biodiversity, it was the relationship formed between these young people from the local area and the businesses that occupied the building.

I often say Global Generation's legacy is not the things that have been created, it is the stories that have been told about what has been done. I share the stories in this chapter because they have become signature stories for an organisational mythology that is about people working together to open up new and unexpected possibilities. These stories are also about the potency of breathing new life into things that are often deemed past their use-by date. This is more than the act of recycling to avoid more environmental devastation. Each of the elements we worked with to create the living roof, the glass, the sewerage pellets, the logs, and branches, carried a story. In her book *Reinhabiting Reality*, Freya Mathews argues that one of the fallouts of modernity is the loss of story. In celebration of manufacturing the new, the earth and everything else is ground down to a neutral substrate in which the "spirit" is lost. In one of the new developments where Global Generation works, I recently read an advertising hoarding for the sale of luxury apartments that read, "In Celebration of Modernism" and in many ways this is an accurate description. These new developments are largely master-planned through cost analysis and probability statistics. Once the shiny and new has arrived, marketeers and events companies are hired in at vast expense to tell new stories in the hope they will create a compelling enough narrative to attract people to come there. Mathews describes another way of developing our future places, which sums up Global Generation's approach.

> When manufacture follows the principle of (natural) fertility, then we will have built a world in which everything, artificial as well as natural, retains its own story, its unique place in the poetic unfolding of the world, and all things are in conversation "in story" with the things around them – just as things are in the "state of nature" … Such recycling of the given into the emergent is not undertaken merely for the sake of sparing further inroads into ecosystems, but out of appreciation for the poetic integrity of existence.[iii]

The summer after the Living Roof was established, there was a terrible drought. In dismay, I watched as wildflowers wilted, grasses became brown and crispy, curled up and finally died. Charlie, the building owner, was losing faith in our endeavours. I was too. One evening, as we pondered the parched roof, Sara Riley, exclaimed in a sudden burst of enthusiasm, "It's all about water, this has happened because we need to find a way to understand and work with water."

Thus emboldened, grant applications were written and a series of phone calls ensued. This led to success with the National Lottery and sponsorship from Thames Water. Rainwater harvesting and a Grey to Green Water recycling system were installed. With the help of grey to green water enthusiast, Chris Shirley-Smith, we ended up with more water on the roof than we could possibly have hoped for. Not only could we create a vegetable garden on the roof, now half the toilets in the building could be flushed with recycled water.

Each individual element – the water, the biodiversity, the energy – had a story to tell. However, the most compelling story was what happened when we brought everyone together to re-plant the roof at the beginning of the summer. For the second time, construction company Costain Laing O'Rourke helped us. They arranged for a 10-ton crane and a road closure to lift several tons of aggregate and water filtration tanks onto the roof. Teenagers and construction guys along with The Office Group Director, Charlie Green, all worked together, raking and sowing seeds. It was like one giant organism that stretched from the pavement to the rooftop. There was a real sense of unity and purpose between us. I overheard one of the teenagers new to our work whisper to his mate, "This is cool – I don't know what it is, I just like the vibe here." During the day, Paul Aiken said,

> I think there's a transformation occurring on this roof. I brought a group of ex-offenders from Camden Jobtrain as we are trying to give them a taste of new pathways. I think that what is beautiful is that with the guys from the Channel Tunnel construction, they are all in flow together, that's the spirit we want to create.

Paul, Sara, and I recognised that the involvement of people of all ages and circumstances is what would keep the spirit of the project alive. Having a vegetable garden on the roof created opportunities for not only growing food but also harvesting and eating together. Around that time, we met an urban food grower, Paul Richens, who by his own admission had been a "worm botherer" since the age of two. Paul would work for Global Generation for many years to come. Soon, Paul was working his magic on The Office Group's roof, and willow baskets were created to grow red and white currants, brussels sprouts, and blue potatoes along with mountain strawberries. The idea of creating transactional relationships through the sale of our own produce began to develop. We discovered that selling a bunch of carrots and a few heads of broccoli provided a simple way for young people to go into places in their neighbourhood that are normally out of bounds. This is how one of the young people involved with us describes the work.

> It allows young people to integrate into places like businesses that we are normally excluded from, meaning that Global Generation works as a medium for young people to emulsify the community together by becoming more environmentally aware; we use nature as the common ground.
>
> Rachel Gates, Generator, 16 years old

The experience of this first rooftop garden fuelled my sense that I wanted young people to have the belief that they could make things happen. I wanted them to be not only pioneers in their imaginations but in the real nuts and bolts, cranes, and concrete world. Amika, a 12-year-old, who took part in the green roof project, put it like this: "I didn't realise this kind of thing can happen on top of buildings … it makes me think of all the things we could do in London."

## Growing with developers

Not long after the living roof was established, I noticed that the proposed King's Cross development included a large number of bio-diverse "living roofs." I was curious and made a call to Argent, the developers. A few weeks later, Roger Madelin, the CEO, met with me at The Office Group roof. I remember dressing up in preparation for our meeting. I managed to dig out some lipstick and a pair of smart black trousers. Roger showed up in his trademark Lycra cycling outfit. It was a good meeting. Roger liked the fact that we were, as he put it, demonstrating that business and activism didn't have to be either ends of the spectrum. He suggested I email him ideas about potential projects, which I did and sometimes he would reply. I also began my own back-up plan. This entailed attending events where I knew Roger would be speaking. I would always catch a word with him in the break. Sometimes stubbornness pays off and this time it did. On one of these occasions, he said to me that he had been thinking about the possibility of growing food on the back of flatbed lorries. A concern for developers and their investors is the risk that a community group might suddenly claim squatters' rights, especially in the case of a garden that has a feeling of permanency. The next morning, I called Paul Richens, who was by then our Gardens Manager. He came up with the idea of growing food in skips (dumpsters) – large enough to climb inside to garden effectively and portable. The idea stuck.

This was all occurring amidst a backdrop of intense financial uncertainty for Global Generation. All of our reserves were depleted and we only had enough money in the bank for the next three months. As a charity, going into deficit is not an option. I knew that if nothing significant happened on the money front, we would go under. My hopes were pinned on a large three-year proposal we had submitted to the Big Lottery, but even then, we would need to secure £50,000 pounds of match-funding within a month. Seemingly out of the blue, at about the same time, I received two phone messages. One informed me, off the record, that we might have good news from the Big Lottery Fund. The other was from Three Hands, a brokering agency that *the Guardian* newspaper had engaged to find a local charity to work with on community-based food growing projects for their new starters. *The Guardian* had recently moved into Kings Place, a new building overlooking the King's Cross Development.

One of the big lessons has been the importance of having an internal sponsor watching our back. At *the Guardian*, this was Carrina Caffney, Commercial Sustainability Manager. I first met with Carrina in an unofficial capacity, as she, like me, was an alumnus of the MSc in Responsibility and Business Practice at Bath University. Carrina advised me on how to make our pitch. She suggested we get the Generators to send in handwritten letters to *the Guardian's* senior management team, which they did:

> One thing you can rely on is our commitment, we are highly motivated generators with young fresh minds and ideas. We believe, coming together, working and interacting with people like the Guardian staff is the best way of making progress. In that way, people will be more motivated, inspired, and will be able to get the idea to the heart and not just the head.
>
> Zak Nur, 16 years old

Paul Richens produced some hand-drawn designs for the first Skip Garden, which together with the end result are in the images on next page. Paul's sketch made everyone smile and say "yes." Armed with the Generators' letters and Paul's drawings, I met with Tim Brooks, the then MD of *the Guardian*; the COO, Derek Gannon; and Viv Taylor, who was responsible for staff learning and development. Two years later, in an interview with one of our Generators, Tim described our first meeting.

> Global Generation turned up with a drawing of a garden in a skip and I laughed – I thought it was funny and brilliant and that we have to do this. So it took about a minute for me to be persuaded. What I loved is that it's like a dream thing but then they were actually serious about making it happen. That combination of dreaming something but being very practical in making it happen was extremely powerful.
>
> Tim Brooks, CEO, Guardian News and Media, January 2011

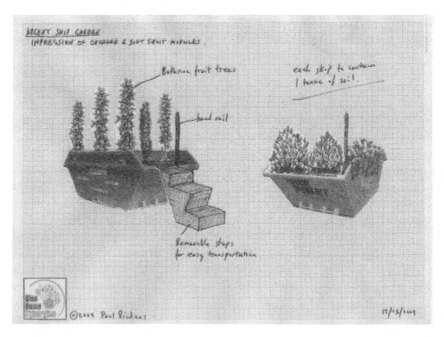

First concept drawing of the Skip Garden presented to Tim Brooks then CEO of *the Guardian* in 2009

Skip Garden Number 1 opposite St Pancras Station – 2009, Photographer Ian Christie

*The Guardian* and Argent gave us the money needed to develop the King's Cross Skip Garden through a process of hands-on business employee and young people's engagement. This provided the necessary match-funding for the Lottery Grant. However, perhaps more significantly in the long term, the combined relationship with *the Guardian* and Argent provided the kind of national platform Global Generation needed to take the work forward, as it brought together talented and passionate people committed to innovation and change in urban regeneration. In both cases, young people and a grassroots can-do spirit were the essential ingredient. This blend of social and physical, inner and outer change has become the hallmark of Global Generation's work.

## Learning from nature

Bumble, carder, and hairy-footed bees feed on wildflowers planted in the midst of the construction. The boundaries of the Skip Garden leak. An older rhythm is heard across the development. If you listen closely amidst the noise of construction and the rise of buildings, a kind of enchantment finds its way through. One of my favourite descriptions of how the natural world has evolved comes from Brian Swimme and Mary Evelyn Tucker in their book, *The Journey of the Universe*. They describe how nature is neither an engineers' blueprint nor is it entirely random; rather, it is profoundly exploratory.[iv] Sadly, this understanding is not so evident in the highly controlled, formulaic approach to nature that has shaped many public places in London: an approach of separation and control that is emblematic of the modernist venture. Global Generation's work with bees has been an important part of our work in supporting children and young people to see nature as their guide. In terms of supporting bees, the most important part of our work is growing food for bees and other pollinators. This has been an integral part of the Skip Garden and we have influenced some of the planting practice across the wider King's Cross Estate, to make it more bee-friendly. Engaging in the practice of bee-keeping has been a way for all of us to appreciate the interconnectedness of the natural world.

### From my Journal – June 17<sup>th</sup> 2011

*It has been a hot Friday afternoon. Brian, our beekeeper, myself, and three young generators have spent several hours lifting frames out of the hives on the rooftop of a King's Cross office building looking for the blue dotted queen. The bees were quiet, the smell was delicious – too delicious? Lots of honey was being made, good stores for the winter but alas it had come in such quantity because there was no brood – no eggs. We thought back and remembered the queen cells we had seen a month before. The original Queen had died and the worker bees were desperately making a new mother. We wondered if any of them had survived. More importantly, had the queen mated?*

*Then we found her; not that different than the rest of the bees but definitely a Queen. However, she was a small Queen which indicated that she had never mated, she had no eggs inside her. She was a virgin queen.*

Our work with bees is thanks to Brian McCallum and Alison Benjamin of Urban Bees[v]. Brian is an expert in all things that are bee related. He is also a wonderful storyteller and he has been an important part of introducing the story of bees and the world they inhabit to the teenagers we work with. For some of them, this has brought about a dramatic shift in attitude. Fifteen-year-old Jihaan was one of our first young bee-keepers and this is how she described her experience of the Friday afternoon we spent with the bees.

> I haven't been here for three weeks but when I came back, the beehives have really changed. It is quite amazing how different it looks. It is like watching a child growing up into an adult. I have changed too; I have more confidence than I did before.
>
> I was surprised that the bees had made a lot of honey since I was gone and that new bees are coming up and old ones dying and there are two virgin queens in one hive. When I started with the bees, I was really scared, I was always in the corner. Now they can be around me and I don't move and I can control my nerves, normally I would be yelling and screaming, I now know how to move calmly so they leave me alone.
>
> As a young person, it is really benefiting me to learn to be brave. One day, if I have children, I know I will be able to back them up and encourage them not to be scared of insects, not to be scared of life.

Global Generation's gardens are noticeably different from the straight-lined precision of the surrounding glass and steel buildings. There is magic in the juxtaposition of the relative freedom of our gardens with the lock stock feel of the city. Using the spirit of nature as a foundational starting point runs counter to many modern design processes, where large areas of city developments are designed through computer-generated programmes. This means that huge swaths of the city have buildings and parks which look more or less the same, a situation compounded by the expectations and rigid parameters of the public planning system. As the well-known urbanist Jane Jacobs describes, since the 1950s "we have been using statistical and probability techniques to create formidable and impressive planning surveys for cities … It then became possible to map out master plans for a statistical city, and people take these more seriously, for we are all accustomed to believe that maps and reality are necessarily related, or that if they are not, we can make them so by altering reality."[vi]

Jacobs goes on to describe how in the 18th century a curious thing happened, the long finger of which I encounter in the places where Global Generation works. We began to sentimentalise nature. We conjured up images of nature scenes imbued with nobility, purity, and beneficence. We then re-created a manicured replication of nature in the city. This superficial sense of nature has a dark underbelly. I have heard developers, in a bid to appear green, say, "We will plant the biggest trees possible." And they do, winging them in by crane and even helicopter across cityscapes. Controlled planting offers nothing to slow the loss of diversity and broad scale pollution of rivers, soils, and seas both within and beyond the boundaries of the city.

Even though sirens, car alarms, and the thrum of traffic provide a constant backdrop, life still manages to push through the cracks in the concrete. The starlings find homes in the buddleia and the foxes still come out at night. A verge between the road and the concrete escaped the planner's sword; ragwort, shepherd's purse, and poppy ... chamomile, wild carrot, and burdock find their way through. These are the untamed areas that have not yet been "developed." Nearby, a new park has been installed, which is "perfect" in its block planting of manicured flowers and leaf shapes; however, an ecologist's report testifies that few creatures live there.

In King's Cross, the evolution of the Skip Garden has been something of an enigma. In a development that prides itself on the delivery of stylish walkways, parks, and gardens, the seeming disorder of a garden that has laid itself open to more spontaneous realities has created a tension for those who are charged with designing and delivering an orderly public realm in this new part of London. Much discussion over the years has been had with landscape architects who govern the design of the open spaces in the development. They have stressed the importance of suitable screened fencing to demarcate the boundary between the manicured and the maverick. The idea of involving children and young people in the co-creation of food-growing gardens in new public spaces has sometimes been met with a tepid reception. Fortunately, theirs are not the only voices amongst the developer team that have prevailed. The Skip Garden has been quietly encouraged by some quarters, who fortunately have influence, to have leaky margins. The nettle and dandelion that flourish outside our front gate express our fondness for the leaks.

Along with gardening, there is also a social artistry to Global Generation's work, which has been key to our success. This involves understanding, empathising, and finding openings within the worldviews of those we work alongside. In practical terms, this has meant learning to be multi-lingual. For example, I soon learned that short sentences and three-line emails might have a chance of eliciting a response from a construction company who work in hazardous situations under enormous pressure. Whereas a more lengthy, process-oriented description was quite acceptable in communicating with the staff at Central St Martins, the art school who are also resident on the King's Cross Estate. Jean Houston describes "how a social artist is someone who can enter into different cultures and organisations and cross the great divide of otherness by having leaky margins to other people's beliefs, style, and ways of knowing".[vii] Over the years local construction companies helped us in a multitude of ways, including moving the Skip Garden to three different locations on the King's Cross Estate, as shown in the photograph on next page.

Skips on the move in King's Cross 2012 – Photographer John Sturrock

## Skip Garden number three

The first and second moves of the Skip Garden were relatively easy; we only had a few skips and knew where we were going well in advance. The idea was that as different areas of the development needed to be built on, the Skip Garden would move around. It sounds simple and makes for a good story. However, as time has gone by, finding space for the Skip Garden has become progressively harder, and as we grew into a larger organisation, we had staff whose jobs depended on us finding a new home. This called upon a combination of trust and persistence and most of all, a solid network of relationships. In 2015, the clock was ticking. We needed to find a new home and where that was to be was not yet confirmed, let alone the finances to begin all over again. An email came, and then another, and finally there was a knock on the door of the little green garden shed that served as our office. This was Julia King from the UCL Bartlett School of Architecture, who had an idea to involve architecture students in the build of the next Skip Garden. I was wary about the challenges of compliance required for an accredited university curriculum. However, I appreciated Julia's persistence and it wasn't long before we agreed to a collaborative project with The Bartlett, led by Julia and her colleague, Jan Kattein. Fortunately, the location was confirmed (or so we thought) but that was all. We still had no money to put into the move. I wrote following blog out of diary entries at the time. It was written in the heat of the build of our current Skip Garden in 2015.

### From my Journal – June 11th, 2015

*I woke this morning thinking about how we might get hold of 30 cubic metres of subsoil for the rammed earth wall that will form the back of our new passively heated green house. We are well and truly in the midst of the moment we all knew was coming … The Skip Garden is on the move. The design of the garden structures and the spaces are a step change on what we have created previously. They include a meeting space on top of a shipping container surrounded by sash windows to create a giant glass house, a chicken coop with a retired silver birch from Hampstead Heath in the middle, a wetland dining area powered by the grey water from the kitchen. To top it all off will be a giant gramophone horn to connect to our bee-hives, bringing the sound of the bees to the rest of garden. All wonderful but … how on earth am I going to get hold of two free lorry loads full of soil?*

A few days later, I sat around a table with five employees from one of the onsite construction companies. We were having lunch before our biweekly lunch and learning session began – "I have a quick question" I said – does anyone know how I could source 30 cubic metres of subsoil. A project director from Bam Construct spoke up, "As we speak, we are pulling out King's Cross sub-soil where our new building will go up, it is just above the contamination line." That was all that was said, the next day I received an email to say that the lorry had been arranged and asking when we would like delivery … job done, well, almost.

The day before, I was worried about the concrete pads for the larger build-ings which included a two-story portable cabin for our Skip Garden Kitchen. "It's sorted," Alistair Mitchell from Carillion quietly said, and then added, "That's £6,000 pounds worth of concrete for just one of the pads." Meanwhile, Kier construction company were working on a complex transport plan which involved a crane lift for the portable cabins and of course the moving of the skips. The students were hard at work in the Bartlett school yard, pre-fabricat-ing many of the structures in readiness for when we eventually would sign the lease on the new site – a site which had been changed three times in as many weeks. In the background, last but not least, our lawyer, Ben Jones from the law firm Freeths was working up (on a pro bono basis) our lease and a set of agreements which would enable the various institutions, especially a univer-sity like UCL to open the door a little wider for the future of practice-based education.

Friday, the 29th of June, was D-Day, and the invitation went out to our friends and through our networks for the opening of the new Skip Garden, which was to be included as part of the Bartlett School degree show and the

London Architecture Festival. This was also the day we needed to move out of our old Skip Garden site and so that evening, a large van full of leftover materials would be driven down our campsite on Pertwood Farm for a much-needed volunteering weekend. On the Thursday before, I pick my way carefully across the site of the new Skip Garden and my heart sinks. There is rubble and sawn-up ends of timber everywhere. The skips for taking the rubbish away – kindly organised and donated by one of our construction friends had not yet arrived. The toilet block made of railway sleepers and scaffold boards looked amazing, but the plumbing wasn't working, and the main window of our café was still boarded up. A familiar thought went through my mind. What will happen this time?

Well, the right people did appear at the right time. By Friday morning, the garden was teeming with people, volunteers in all directions – university students, young people doing work experience, and local residents. By 2 pm, the advertised opening time, the site was clear of rubbish and the last of the signs were being hung. Last but not least, a cocktail bar was operating from one of the garden skips. The students from the Bartlett, who had worked long and hard, were beaming as visitors flooded in to see what had been created in a relatively short period of time, with 90% reclaimed materials and a start-ing point of only £500 budget for each structure. This set of photographs show; one of the many workdays that went into creating this more ambitious iteration of the Skip Garden, some of the innovative structures the students designed and built and the wonderful end result of a sometimes hair-raising process.

Making of the Skip Garden Number 3 – with Jan Kattein, Dave Eland and Students from the Bartlett School of Architecture – photographer Jane Riddiford

Skip Garden Office and Skip Garden Café Cool Store, 2016 – photographer John Sturrock

Skip Garden Chicken House, 2016 – photographer John Sturrock

Skip Garden Up and Running with the Skip Garden café manager Fi Doran
photographer John Sturrock

Along with the students and their tutors, Jan and Julia, special thanks go
to the site manager, Dave Eland. Dave is an architect who spent many years
running self-build projects in Eastern Europe, so he was made for the job. He
worked tirelessly, supporting the students, and designing and building the classroom, kitchen, and cafe for Global Generation's teaching and income-generation
activities. He has involved volunteers of all ages and stages of experience. This
has included groups from the Guardian, Google, and local residents, along with
primary school and secondary school pupils. As the build was happening, project
lead, Jan Kattein, whose PhD research focussed on collaborative architecture,
described how the Garden provided a very real alternative to how architecture
training is typically done – a process which embraced the bumps as much as the
breakthroughs. This is how he describes the way we worked together with lots
of different players to build the Skip Garden:

> For the first time in living memory, Bartlett students designed and
> built a real project on a real site for a real client. The project sets a
> precedent for how architecture can be taught, but it also sets a prec-
> edent for the role of architectural practice to engage and empower
> communities. Engaging in detail with the charity's ecological and
> educational remit has brought about designs that are unique, specific,
> and responsive. Almost all materials are reclaimed, many of them from
> the King's Cross development sites surrounding the garden. Earth dug

out adjacent to the site was used to construct a rammed earth wall. Reclaimed scaffolding boards form the filtration beds for the grey-water recycling scape, a reclaimed shipping container serves as foundation and base for the twilight gardening space, and a produce cool store is built from earth-bag walls made from coffee sacks donated by a local roaster. Engineering firms are giving their time to help with designs and calculations. Large construction companies and the City of London are donating materials. Developers and local businesses are providing funding. The local authority is advising on planning restrictions. The urge to be part of a larger whole is galvanising a community of supporters and contributors who are keen to advance their profession and question the status quo.

We are proposing a distinct departure from the polished appearance of university portfolios to reveal the contradictions, illogicalities, and un-predictabilities that shape the architectural design process in the real world. Students are encouraged to record mistakes as well as successes and reflect on endorsement as well as criticism. The research log treats failure as an integral part of the learning experience. The greatest problem of our contemporary risk-averse society is the disrepute associated with failure; whereas in reality, the recognition, analysis, and dissemination of failure in an academic pursuit are essential constituents of the learning experience.[viii]

## Reflections

Along with the successes, there were failures; the cool store leaked and the earth bags didn't hold, the chicken house, whilst beautiful, was hard to clean and the greywater filtration beds overwhelmed with the volume of water from our café and became anaerobic and smelly. The failures tested us and invited us to creatively engage with the next steps of taking the garden forward. Structures in the Skip Garden morphed and changed; a water catchment roof became a living green roof and the greywater system a large herb garden. The sense of exploration and ongoing evolution of the gardens means that not only the plants but also the built structures are what creates a living garden. Whilst I often feel pressurised by the assumed predictability of the master planning process, I have learned to duck questions like "what does success look like to you?" In my view, if we knew what success looked like, it wouldn't be successful. We have deliberately entered into projects in which the end result is not yet written and have learned to travel hopefully. On many occasions, breakthroughs and unimagined possibilities have come out of bumpy, difficult ground. Living through struggles means that in each of our gardens, I have been able to proudly say, "This is an exploratory space where some things worked and some things didn't."

I learnt over and over again that understanding the currency of relationships with a diverse range of stakeholders is vital in a garden with no permanent abode and an organisation committed to community building. It was important that we understood and at least to some degree empathised with the pressures and concerns that different people were embedded in. For example, a concern for developers and their investors is the risk that a community group might suddenly claim squatters' rights. Our path has not been one of protest and civil disobedience, but rather a gentle form of activism which has meant creating change by inviting people in to eat, to garden, and to dialogue, and in this way slowly bringing them on board. This has created a web of connection in the gardens which is about the plants and insects and it is very much about people and the relationships between people. The involvement of people of all ages and circumstances is what has kept the spirit of the project alive. Beneath the straight jacket of business as usual, we have discovered a well-spring of good-will in most people. This is made all the more accessible by giving children and young people opportunities to have a voice, which in turn has had a humanising effect on others. Working with recycled materials provided excuses to involve different people in the garden. Everything our benefactors provided, whether it be the earth for the rammed wall, the six kilometres of scaffold boards, or the sash windows, carried a story about where the materials came from. These stories, even if they weren't explicitly known, were part of growing the atmosphere of the gardens.

In working with large companies, it has been vital to have internal champions who could not only speak for the work, but who were prepared to have offline conversations about how to navigate what can sometimes seem like an impenetrable system. In that sense, I have often said we don't work with organisations, we work with people.

## Notes

i  Mathews, F. (2005). *Reinhabiting Reality:Towards a Recovery of Culture*. Albany: State University of New York Press, p. 200.
ii  King, U. (2008). *The Search for Spirituality – Our Global Quest for a Spiritual Life*. New York, NY: BlueBridge, p. 44.
iii  Mathews, F. (2005). Ibid/*as above*.
iv  Swimme, B. and Tucker, M.E. (2011). *Journey of the Universe*. New Haven, CT: Yale University Press, pp. 52–53.The book and film by the same name have strongly influenced Global Generations Cosmic education practices. Mary Evelyn ran a workshop in the Skip Garden in 2015 and has remained a keen supporter of our work ever since.
v  Brian McCallum and Alison Benjamin, are co-authors of The Good Bee and Bees in the City, run this website https://www.urbanbees.co.uk/.
vi  Jacobs, J. (2005). *The Death and Life of Great American Cities*. New York: Random House, p. 438.
vii  Houston, J. (2013). *The Urgent Need for Transformational Storytelling*. Online [https://www.youtube.com/watch?v=5Yqo1qc-TLY] accessed 15.2. 2020.
viii  Kattein, J. (2015). *Made in Architecture: Education as collaborative practice*. *Arq (215)* 19.3. Cambridge University Press 2016, pp. 295–306.

# 3

# LEADING AS A WAY OF BEING

Leadership should be aimed at helping to free people from oppressive structures, practices and habits encountered in societies and institutions, as well as within the shady recesses of ourselves.

Amanda Sinclair[i]

This chapter takes a closer look at some of the ground and the glue behind the community green spaces that Global Generation has established. A north star for our educational approach and our community regeneration work has always been an active engagement with the question of how are we as individuals and as an organisation. Are we growing a sense of community amongst ourselves as a team? Are we expressing the kind of values and ways of work that we hope to engender in others? When I am learning about other organisations, one of my first questions is how they are themselves as a community. In the following pages, I describe my initial aspirations for how I wanted us to work collaboratively and creatively in Global Generation. Perhaps most significantly, I share illustrative accounts of what my experience was like amidst the messy day-to-day reality of actually doing work on the ground; a formative reality which sometimes fell short of my aspirations and at other times exceeded them. These, and many other stories, are why I often say a key component of leading well and growing community is the ability to embrace one's own "bumpy ground" as fertile territory for learning and changing. These stories are about meeting obstacles, reflecting on actions and stretching into new ways of being.

## An awkward word

Leadership is a word frequently spoken of and written about. It is a word that I grappled with during the five years of my doctoral inquiry which focussed on my experience of leading Global Generation. In academic terms, I have earned the right to be a leadership scholar; however, I have never felt completely at home with that word. When I hear people speak about leadership or when I say it myself, I sometimes feel an inward contraction, as if it is referring to the hallowed halls of power and privilege. The place that is for generals and captains or at least for those who have been trained in some kind of secretive art. I either start trying to envisage something that perhaps should never be pinned down or I feel sceptical about the large sums of money that are exchanged in return for training in leadership. I am not alone in my reluctance to join the ranks of leadership. Amanda Sinclair,[ii] who was professor of diversity and change at Melbourne Business School, describes how in her interviews with woman leaders, she came across many who refused to put themselves in the leadership category. For some, it was modesty and for others, there is as an active rejection of a territory dripping in white patriarchy. I think of my mother, who even in her 90s, is very much the leader in our family; however, she has never called herself a leader. Such is the reification and cultural potency of the leadership business that it is easy to overlook the very simple fact that in one way or another, many of us have experience of leading and have plenty of common sense about what makes good leadership. Leading is part of what it means to be human. I think most people would agree that leading well is an important part of growing positive change in the world. As I see it, leading is not so much about the things we do, it is how we are that is important. In that sense, leading is a way of being.

Despite the vast body of literature, contemporary scholars have recognised the difficulty, and some would say impossibility, of studying the invisible phenomenon of leadership. Consequently, very few researchers agree on what leadership actually is.[iii] As already mentioned, I feel that a useful addition to the large canon of leadership literature written by disembodied authors about "other" leaders would be more first-person, auto-ethnographical accounts of the lived experience of leading. I realised that a big part of my resistance to calling myself a leader came from the connotations of individual drive and authority associated with the word leadership, which clashed with my intentions for Global Generation to be a collaborative organisation. In 2003, in the early days of shaping Global Generation, I encountered the work of Brian Swimme and Thomas Berry and had the privilege of hearing them speak in the same year. I was inspired by the way Swimme and Berry imbued ecology and cosmology with meaning, values, and a sense of purpose which seemed directly applicable to my aspirations for Global Generation.

> There is a deep pull within us to discover and be true to the fact of our own interrelatedness. It is a way of creating a bigger human being. Our identity comes out of this place of togetherness, we must become kin, internally related.[iv]
>
> Swimme and Berry

In my diary at the time, I wrote of developing an organisation that would work like an ecosystem, with balance and co-operation, homeostasis, and dynamism. The ecological metaphor of an organisation operating as an eco-system spoke to me of collaboration. I soon learnt that it is not that simple. Experience showed me that leading is not a fixed way of being; it is both an individual and a collective process. More crucial than whether it is an individual or a collective process is having the awareness to respond appropriately to whatever is required in any given moment. Whilst Global Generation has been the work of many hands, it has also been about individual drive and risk-taking. Getting projects off the ground often called for precociousness, persistence, and downright stubbornness. This was especially true in the early days when I had neither reputation to call upon, nor the confidence and experience of how to let go and make room for the ideas of others.

## Together and alone

Our first significant London project was the Living Roof Project on the roof of a King's Cross Office Building, described in the previous chapter. I wanted to use logs from Hampstead Heath to recreate a feeling of the countryside in the city. This was met with disbelief by one of Global Generation's co-founders who was an old friend. She firmly said, "You can't do that." There was the sense from her that I didn't know the rules. Rebelliousness rose inside me; "Yes we can," I asserted. I was aware of speaking with undue force; I wanted to prove it was possible. Three weeks later, the Hampstead Heath conservation team were carrying logs and branches onto the roof of the office building, what's more it was all for free. Discussing my behavior with a doctoral colleague many years later, when I was prepared to be more self-critical, she suggested that it showed me to be an obstinate subversive, a mediator of complex systems, a leader of inspiration, a pragmatist, and a power broker. I would not overtly signup to many of these qualities, but I do harbor a secret liking for the term "swashbuckler," so perhaps she was right.

In 2004, I read Leadership and the New Science by Margaret Wheatley. For a long time, it was the only leadership book I owned. I returned to it again and again. Wheatley describes leadership in many organisations as still being based on the Newtonian idea of certainty – that we live in a fixed and knowable universe:

> As the earth circled the sun (just like clockwork), we grew assured of the role of determinism and prediction. We absorbed expectations of predictability into our very beings. And we organised work and knowledge based on our belief about this predictable universe.
>
> Margaret Wheatley[v]

I can remember literally kissing the pages, thrilled by Wheatley's account of leadership in an organisation based on our more recent understanding of an evolving universe in which the future is not yet written. To put these ideas into practice was challenging on a number of fronts. Operating within the public sphere without being constrained by it often felt like a multi-lingual exercise. For starters, there was a need from our funders for fixed plans. Contradictions were implicit in the process. On one hand, we had to promise set outcomes and outputs and, on the other hand, communicate that project plans would evolve through the relationship between everyone involved. Furthermore, I felt different and conflicting drivers in myself, which was sometimes confusing for those I was involved with. I was interested in what would emerge through the relationship with others. However, at times, I felt a burning, individual drive to make things happen, which I found hard to define or even speak about. For longer than I care to remember, establishing Global Generation was about spreadsheets and sleepless nights as I was constantly worried about funding. My insecurity reinforced the feeling that I needed to make things happen on my own. This hindered the growth of a shared sense of responsibility and I fell-out badly with one of the original team over this.

In this process and in many of the stories shared in this book, I acted as a fire-starter, with an impulse and a vague outline of something that I felt was wanting to happen. As time went on and the Global Generation team expanded, others also followed their passions and experience, sparking new projects and new possibilities for the young people we work with. Actually shaping and delivering good work in this un-fixed context was definitely not a solo affair; it was ensemble work, which developed through everyone's participation. In this regard, leading has been about storytelling and dialogue in action. Physicist, David Bohm, explored the potentials of collective intelligence with large groups of people. Bohm's description comes close to my own experience of working in a small ensemble.

> It's not like a mob where the collective mind takes over. It is something between the individual and the collective. It can move between them. It's a harmony of the individual and collective, in which the whole constantly moves towards coherence, so there is both a collective mind and an individual mind, and like a stream, the flow moves between them.
>
> Bohm (1996)[vi]

In many regards, I feel I have had a foot in both worlds; following both known and unknown routes. Like most people who run small charities, I have worried about budgets and fulfilling promised outcomes to ensure we had money to take action. On the other hand, in setting up Global Generation and still now, I experience an overriding pull towards something that has no pre-determined outcome. It is close to Patricia's Shaw's[vii] description of ensemble work. This is a kind of opportunistic improvisation which requires moving into what might be emerging, without too fixed an idea of what each move will lead to. Our chair of Trustees, who had a career running an international trust company for a major global bank, described why he was attracted to join Global Generation:

The real attraction is something almost unwritten. GG has a plan of where it is going over the next few years, but this is not set in stone. It will be changed and modified, improved upon and reworked by everyone involved in some way ... AND THAT's NOT JUST OK, IT's WHAT GG IS ABOUT. The end result is not a vision we can see yet. We can set out the framework to get there, but the contributions made during the journey will shape something new and quite unusual, and being part of that changing and uncertain future is exciting. Few organisations are brave enough to set out not knowing exactly where they will end up, and how others will influence them on that journey, but we will.

Tony Buckland, July 10th, 2012.

As a young woman working in an outdoor theatre company, I came across theatre director, Peter Brook. Years later, some of Brook's descriptions of theatre direction resonated with my experiences of leading. "The director will see that ... however much home-work he does, he cannot fully understand a play by himself. Whatever ideas he brings on the first day, must evolve continually, thanks to the process he is going through with the actors.[viii] In the same way, my sense of vision doesn't feel personal to me. I often say vision is what grows in the footsteps of shared commitment. When my colleagues and I stop and make space to connect with the fact that we are not fixed entities but part of the living breathing process of nature and the wider universe, we begin to draw on the same ground. This is when I feel trust and confidence that the work of our team is going to be good. It is when I let go. I find this process pregnant with purpose, confident but not in a concrete or certain way. Seeing ourselves and what we are part of as a vast interconnected, evolving "story" provides the imaginative confidence to step beyond what we already know.

## Patrolling boundaries

In the re-telling of events, it is easy to convey a "smooth story." From a practice point of view, it can make the work sound all too easy. It was thanks to my doctoral supervisors that I became interested in how I embody the gritty ups and downs of my experience of leading. They noticed that I never mentioned the word leadership and that I tended to focus on things going well rather than conflict. One of the faculty, Chris Seeley, busted me on this point by saying, "It is all very nice what you are describing; however, it is clear that you are patrolling boundaries with an iron fist." Chris was the examiner of my transfer past the MPhil stage of the doctorate. In order for the transfer to be agreed, she asked me to write an additional 10,000 words on the times I have needed to be dictatorial.

In the Skip Garden, the strongest boundaries are not set by walls and security gates but by a tacit sense of ethos. After Chris's challenge to me, I started to pay attention to how I decided what would pass through the Skip Garden boundary.

Sure enough, a situation soon occurred when I needed to put the barriers up. An investment bank was about to pay for their graduates to spend a day with us in the Skip Garden. An email arrived from their Learning and Development Manager asking whether I would mind, due to dietary needs, if they ordered take-out pizza for the graduates instead of eating our home-cooked lunch made from Skip Garden produce. It came down to a strong gut-feeling for me. As I read the email, I said to my colleagues: "If they don't want to enter into the spirit of what we are trying to do, I don't want them with us, regardless of the money." Understandably, the bank had no prior context with which to relate to the work in the Skip Garden. It took a number of frustrating emails and a final phone call to reach an agreement.

ME: "There are a few key elements that have gone into creating a space in the middle of a construction site that enables a positive atmosphere between people to flourish. As we are a food-growing site, cooking and eating together is a core part of our ethos. Stepping beyond the familiar business as usual offered by the high street is another important ingredient. In another context, I would have no problem with you bringing take-out pizzas. However, in terms of keeping Global Generation's foundation strong, I am not able to agree to your request."

LEARNING AND DEVELOPMENT MANAGER FOR THE BANK: "Thank you for coming back to me. I understand your view. As a compromise, would it be ok for us to have the pizza delivered and to eat food outside rather then bring it on-site. Would this work?"

ME (TO THE GLOBAL GENERATION TEAM): "No!!"

As the emails went back and forth, I asked the rest of the team what they thought about it. "I am with you a 1000 percent if you cancel the day," said Paul Richens, our Gardens Manager, "If you compromise on our ethos, we have nothing."

Eventually the bank agreed to come on our terms and we all cooked lunch together.

## What do others think?

As well as wanting to know about the ways in which I was dictatorial, I also wanted to know about the effect I had on others. I asked my colleague, Nicole, about her experience of how I was leading. At that point, Nicole was our Youth Programmes Manager. She and I had been working together for three years and we had seen the growth of Global Generation from three people to ten. Remembering fallouts in the early days of the organisation, I was anxious how I would respond to what might need to be said. Eventually, I plucked up courage to broach the subject: "As part of my doctoral research and for what we are doing in general, it would be good to find a time to speak about how you experience my leadership." Nicole said she would think about it.

A few days after my question, Nicole said, "I have been thinking more about your leadership and where I notice it. One of the main things is that you are always on the ball about the methodology we use. When other people lead aspects of our educational approach that you have developed you are strict in a good way. For example, you insist that we should always book-end our sessions with reflective practice. You trust us to go out and do the work and give space for us to develop our own way of doing it, with your guidelines being there, otherwise it wouldn't be Global Generation's work."

I felt a combination of appreciation and reticence about Nicole's reference to Global Generation's work. How is Global Generation's work defined, and by who? Is there a danger that particular methods could become fixed and limiting? Her comment about being "strict in a good way" was to return to me over the months to come. I knew that I can be definite and I was curious about the effect this had; did it shut people down or did it provide helpful boundaries? "How do you experience my insistence, or even dictatorship, of certain principles?" I asked. Nicole said: "In any conversation you have with anyone new, you speak about the deeper side of the work, like in our interviews with people applying for jobs, and it is in everything you write. The other way you do it, which is becoming more deliberate, is to create a space for people to think about these ideas for themselves, like the meditation and the engaged philosophy sessions and supporting us to do other training."

It was good to hear Nicole speak about the value of investing in the staff team's own learning, which was a new step for the organisation, and I said to her, "I have seen how the work has flourished when I have had the chance to step back and think more abstractly. So, I wondered what would happen if the whole team were given that space."

I appreciated Nicole's comments and I was aware that the question "how do you experience my leadership" is not simple. Power dynamics can be subtle and hidden and can get in the way of an open discussion. The nature of my role as founding director meant that I held more power than Nicole who at the time was our Youth manager. Even though I shared this account with Nicole, I wondered if she felt free to talk to me, without the constraints of me being her boss. Was I subtly eliciting from her what I wanted to hear? How much did I selectively hear and report on our conversation? Was there a forum we could set up where the power dynamics of job hierarchy would fall away?

## A difficult conversation

"Have you got a minute?" I said to one of our young interns, who I will call Walter. The frosty morning was showing a rare glimpse of sun and everything seemed to be shining. The conversation was unplanned, the moment felt right, to me at least, and I took my opportunity. "It might be good if someone else can do the garden tour for the visitors today, so you can come to the lunch time meditation session we are having. I think a next step for you is to go beyond what you already know. You are very bright and pick things up fast, which is great, and there's more. For example, I've noticed that you find it difficult in engaged philosophy," I said.

"Yes, that's right," Walter said. "I struggle with it."

I then went on to say, "It makes sense; it's not always easy for any of us, but there is a real potential if we can learn to inquire together with others and through doing that step into new territory. That is why the focus is on listening and building on what others have said, rather than just bringing our already pre-formed conclusions."

"The thing is," said Walter, who was a 23-year-old Masters graduate, "I have had a very fortunate life and I am lucky enough to feel that I don't need to develop. I am happy with how I am."

This was getting challenging. "There is a vibrancy that comes when we stretch ourselves; like plants growing we become more alive," I said.

"I don't think I will develop by exploring philosophy; I need to be given responsibility and chances to go out and do things on my own," replied Walter.

At this point, I tried to make our discussion less personal. "There can be limitations for an organisation if everybody just goes around doing their own thing – we could all too easily become a group of separate individuals bumping into each other," I said.

In hindsight, I could see that I was saying and implying a lot to Walter. Reflecting on Nicole's comment to me, I would say this is an example of where I was "being strict" and not in a good way. I hadn't really listened to why Walter was uncomfortable or respected what he might need in that moment.

Several hours later, Paul, our Gardens Manager, came to me distressed. He explained I have an unhappy guy working with me today. Walter said you told him he wasn't alive in his communication. He doesn't like the engaged philosophy sessions. Is there a way he can step out of them with dignity? He is scared because he wants to go for the job Global Generation is advertising and he doesn't want to ruin his chances."

My head start to spin, there was a lot going on especially as the whole interaction with Walter touched on aspects of our work that I had struggled to bring to life over many years. I was not willing to let it go and I realized that I needed to be more careful and less impulsive in the way I spoke to Walter. I had been harboring the thought that in time Walter would take on more responsibility in Global Generation, which carried expectations on my part. I wanted to know what our education team had to say about the situation.

"Do you think it is important for people involved in the educational side of our work to come to engaged philosophy?" I asked.

"Absolutely," Silvia said, stating that it is where we can really inquire into what we are doing. That she was more adamant than me was encouraging and also surprising. It highlighted to me that despite the fact that I have had an inner conviction about how things should be done, I have been tentative about verbalising this too strongly.

Afterwards, I wondered if I gave enough room for Paul to express his concerns rather than plying him with others' reinforcement of my actions. One of the things Walter had said to Paul was that he did not like the fact that he didn't feel he could disagree. I said to Paul that in engaged philosophy we are exploring particular dialogue practices, like following a thread, that we are introducing to the young people involved with us. In school, they learn to debate and disagree, but, in my experience, they do not always learn to really listen. Paul has not taken part in engaged philosophy and so I suggested that he and Walter speak to Silvia and others, as I was too invested in the idea that as many of the Global Generation team as possible should participate.

Walter's comment "I don't like the fact that I can't disagree" stayed with me – was there truth to what he was saying? Would it help the process of growing a shared sense of purpose if I made more room inside myself for disagreement? Did I run the risk of surrounding myself with people who simply agreed with me? One of my mentors, whom I will introduce later, has been Mary Evelyn Tucker. I thought of her description of how evolutionary cosmology re-orients us from the fixed idea of "other-worldly perfection and goodness" to a quest toward "participation in the dynamic evolutionary process."[ix] In this light, diversity and dealing in difference is the stuff of the Universe! I felt a gap between my aspirations and what was actually happening on the ground. I noticed that I wanted to stay with my discomfort and keep these questions open; rather than staying with any defensiveness.

## Shifting values

Between the interaction I had with Walter, described above, and the next engaged philosophy session, I spent three weeks in New Zealand. During that time, I began exploring the way my forebears related to the land, described in the first chapter. This made me particularly aware of the force in my family history to advocate one's own plan. Engendering a pioneering spirit has been at the heart of so much I have done in London. I wanted young people to believe they could make new things happen. Whilst I was in New Zealand, I became curious about shifts in what I valued. In a conversation with a school leader, I found myself habitually reaching for the word *pioneer;* I noticed that I did not want to grasp it; the glamour had gone. I felt the raw bullishness, the taking that could strip the land bare, generation after generation. I was curious to see who I was beyond this side of me. I became interested in other words in my lexicon, the juxtaposition between pioneering and development. Changing on the outside or growing on the inside, control or curate?

When in New Zealand, in the Wairarapa, where I spent my teenage years, I often have a ritual morning swim in the Ruamahanga river. On the way to the river, I run through the wet grass following the patterns in the ground. One year on a trip to New Zealand, I had bought a small locket and filled it with soil from the edge of the river. It was a reminder of where I come from. Over the 12 months between visits, the soil had washed out and the locket was empty. I intended to fill it again. At the end of my visit, wheeling a luggage trolley through the native bush planting between the domestic and international terminals at Auckland airport, drinking in a last look at the intense blue sky, a moment of panic filled my mind; I had lost my ground. I felt my locket and realized I had left it empty. The airport soil was no substitute for the Ruamahanga soil. As I contemplated what the locket might now mean to me, I felt myself fall into a different ground, an empty ground. I decided to leave the locket empty, as a reminder that a way forward from the intense emotions I had experienced in relationship to my history would grow through listening from an empty space of not knowing inside myself. It is a space that existed before history, before anything ever happened. It is the dimension of life Māori call Te Kore.[x]

Considering my cultural conditioning provided a pause, particularly in regard to my contribution in a GG engaged philosophy session, which had been arranged by a colleague for my first day back at work. The session was to be an exploration of our individual values and those of Global Generation and where they might meet. After the interaction some weeks before, I was concerned about how it would be between Walter and me and so at the start of the session, I reminded everyone (but mostly myself) of guidelines for encouraging dialogue: listen deeply and stretch ourselves beyond what we already know, to give room for new understanding to emerge between us.

As the light faded, we sat together in the Skip Garden. At first, the conversation was awkward and then a spark took hold and I felt the thrill of ideas seemingly rising up by themselves. Walter was saying something that took our attention, especially because it was coming from him. "As an organisation, we value keeping an inquiry alive about what we are doing," he said.

"Yes, exactly. It's not about coming from fixed ideas about what we should be doing, but rather it's about being open to what is happening and then being ready to respond," said Rod. Then everyone started to speak with their take on what was developing between us tumbling in. Together, we described the ways in which good dialogue can be helpful in going beyond more superficial differences. As each person spoke, I felt myself relax into a current of energy that seemed to go beyond any one individual.

… This is how the evolutionary process works in values … the inquiry is like bamboo sticks bending in the wind … questioning things and changing and adapting our ideas is making us stronger … the space feels natural and organic … there is a freedom to make changes in our own time, in our own way.

As I saw the energy spreading, I felt my attention shift from what I wanted to say to what was coming out between us. Silvia said, "Jane, you and Rod have names for what we are doing, you sometimes speak about the universe's story and evolutionary philosophy. I cannot call myself a follower of this philosophy, but it is important that we have these opportunities to speak together and it doesn't really matter what our different influences are."

After about half an hour, Walter said that it was interesting that even though we were meant to talk about Global Generation's values and no one had actually named them. It seemed to really land with him, that we might all hold different values and be coming from different places, but the fact that an exploration was happening was the important thing, rather than having a list of values.

I was excited by what was happening and I felt there was more to say. However I chose not to say anything. I was thinking to myself, it's not that simple. Silvia then said, "It's not totally random though; we do know if a path is not the right one."

Rod named question I had been grappling with for some time, "Can we be strong in our intentions but spacious and open at the same time?"

Making room for diversity and holding purpose continued to be a live issue for my colleagues and I. After reading this account, more than a year after our discussion, Silvia raised important questions about the episode with Walter:

> What a strong memory and what an important moment for me. The episode is still alive in many ways, in different shapes. I wonder how much we can spread our values without preaching. How much are we ready to accept diversity at GG? Is there really a space for everyone's ideas? Are we really working with the whole community? Is the whole GG team asking questions about their role, what are the potential boundaries?
>
> – November 26th, 2014.

After the incident with Walter, I noticed that I still carried a sense of purpose about what I wanted to happen between us and those we engage with at Global Generation, but I carried it more lightly and with more curiosity.

## Reflections

The past is sedimented in the present and I think it needs to be taken into account; this reality has more influence than one might realise. It has been helpful to take a long view on the forces that drive me. This has meant understanding the cultural conditioning within my family background, which has brought more self-awareness and choice within the present. Knowing the limitations of my own behaviour has helped me empathise more with others.

During the dialogue with my colleagues, identifying the colonialist drive to control land and people in my back story helped me resist the compulsion I often experience to talk too much and to control what others might say. In the case of our dialogue, it would have been with fixed and much-cherished ideas about what I thought Global Generation's core values were. By each of us suspending fixed ideas, we managed to think together and find a shared sense of purpose, particularly in terms of a mutual understanding that Global Generation's primary value, at least at that point in time, was inquiry. In his book, which is called, On Dialogue, William Isaacs[xi] writes of the importance of welcoming opposing views, actively making room for difference and opposition. In contrast to Isaacs, my experience was that the core qualities for the kind of dialogue we had, do not develop concurrently. To constructively oppose, I needed to first learn to really listen, both to others and to the subtler dimension of interconnection within a group. Like Isaacs, I have found that the practice of sitting still has helped me suspend fixed ideas and let go, moving into this deeper, more unknown ground from where I could really listen.

It was unrealistic to hold myself to a stringent ideal of always working collaboratively. I have tried be fluid and present within each situation. This means leading has sometimes been a collaborative process and other times, it has been an individual endeavour. I see it as a dance in which the skill is in reading the situation and taking the risk to behave in whatever way is called for, even if that is uncomfortable. Most importantly, it has been about learning to slow down and listen. As subsequent chapters reveal, endeavouring to listen well is an ongoing practice.

I have learnt to be 'multi-lingual,' promising clear outcomes and at the same time engendering confidence in a more unknown path, in which vision, understanding and possibilities emerge out of relationship with others. I view this ensemble work as a form of dialogue in action.

In my experience, it is important to set boundaries and not be ashamed of doing so. Most of the time, they have been invisible but when they are needed, I have been able to step forward and make them clear. The boundaries have been reinforced and felt within a shared ethos. Time and time again, commitment to growing a shared ethos has paid off.

## Notes

i Sinclair, A. (2007). *Leadership for the disillusioned: Moving beyond myths and heroes to leading that liberates.* Crow's Nest, NSW: Allen & Unwin.
ii Ibid
iii Jackson, B. and Parry, K. (2011). *A Very Short, Fairly Interesting and Reasonably Cheap Book about Studying Leadership.* London: Sage and Ladkin, D. (2010). *Rethinking Leadership – A New Look at Old Leadership Questions,* Edward Elgar: Northampton.
iv Swimme, B. and Berry, T. (1992). *The Universe Story - from the Primordial Flaring Forth to the Ecozoic Era: A Celebration of the Unfolding of the Cosmos.* New York, NY: Harper Collins.
v Wheatley, M. (1999). *Leadership and the New Science. Discovering Order in a Chaotic World.* San Francisco, CA: Berrett-Koehler. p.28
vi Bohm, D. (1996). *On Dialogue.* Abingdon: Routeledge. p.31
vii Shaw, P. (2002). *Changing Conversations in Organisations: A complexity approach to change.* London: Routledge. p. 42
viii Peter Brook (Ibid)The Empty Space 1968, 47
ix Tucker, M.E. (1985). *The ecological spirituality of Teilhard.* Chambersburg, PA: Anima Books, p.10.
x Marsden, M. (2003). *The Woven Universe: Selected Writings of Rev. Māori Marsden.* Te Ahukaramu Charles Royal, ed. New Zealand: The Estate of Rev Māori Marsden.
xi Isaacs, W. (1999). *Dialogue: The Art of Thinking Together.* New York, NY: Currency/Doubleday.

# 4

# I, WE, AND THE PLANET

That we should have caused such damage to the entire functioning of the planet Earth in all its major biosystems is obviously the consequence of a deep cultural pathology.

Thomas Berry[i]

Several years ago, I heard a talk by cultural historian Richard Tarnas[ii]. He described how for the last few hundred years, ever since we discovered that the sun was the centre of our solar system, we have been living in an era of solar logic. In other words, more than faith or feeling, the dominant paradigm trusts tangible things that we can see, taste or touch. Many of us have been so entranced by the progress that the observable scientific method has brought that our sense of reality has become overtly materially focused. Not only is this a driver for consumerism; materialism influences how we approach the environment, community-building, and our own lives. Many of us have lost sight of the fact that meaningful work goes beyond the world of things. On more than one occasion, I have witnessed the disappointing results of separating things out and over-privileging the outer tangible dimensions of life. With the best of intentions, a local authority and on another occasion a developer tried to re-create 'Skip Gardens' in the hope that they would magically grow community. Whilst it is an accolade that, at least in London, the Skip Garden has become an emblematic symbol of what a community garden might be, focusing on the Skips alone misses the point. Testimony to this is that there have been at least two lifeless and rather sad-looking 'Skip Gardens' in London. These failed efforts motivated me to write this chapter, which is about the philosophical underpinnings of Global Generation's work. I will endeavor to take you beyond the world of things, into the 'I, We, and the Planet' approach that shapes and brings the life blood to how Global Generation runs school and community workshops and how we run our organisation. Through narrative accounts, I will share examples of community-building practices that have helped children and young people explore the inner, outer, and collective dimensions of who they are and what they are a part of. I will introduce the universal nature of the 'I, We, and the Planet' dimensions of experience, which I discovered in Māori Mythology and in ancient Greek philosophy. This chapter is also an introduction into how storytelling is a way of understanding and expressing intangible dimensions of life that are often hard to articulate because they are beyond the observable world of solar logic; dimensions of life that are, as Richard Tarnas would say, "deeper than daylight can comprehend."

## Action research – a catalyst for change

In 2008, I was introduced to action research on the MSc in Responsibility and Business Practice at Bath University. As mentioned earlier, action research is a qualitative, inquiry-based style of participatory research. First Person, Second Person, and Third Person inquiry are the three territories of action research which can be translated as me, us, and them. This means that rather than researching others out there, in action research, the researcher is also the subject of the inquiry. Whilst writing about experience is an important focus, the main objective is to be involved in bringing about positive practical consequences, in whatever form that may be. Thanks to the process of action research, I began to ask questions about whether or not I was bringing all I had to offer to my work with Global Generation. I was aware that there were things, such as my lifelong interest in meditation and a curiosity about story and ritual, that I seldom brought into the work I did with children and young people. I never spoke about these things in professional contexts. At the same time, I questioned the fact that many environmental organisations are primarily focused on outer change. Some of this is no doubt due to the narrow parameters of various funding and accreditation bodies and an overall expectation for tangible, measurable results. It seemed doubtful to me that growing vegetables, changing light bulbs, and counting carbon, whilst all important steps, were enough to change the world. There was something more about how we see the world that I felt was important. These questions became all the more pressing when working with teenagers who had little connection to environmental concerns or the natural world. Somewhat, tentatively at first, I began to bring more of myself, my interests and experience into my work. In his book, *Synchronicity*, Joseph Jaworski describes his leadership training programme for senior executives. His book gave me the confidence to speak about the inner dynamics of my work. He writes; "I knew I would have to find a way for the leadership forum to give its fellows an 'inner education,' so that they would identify themselves with all humanity, if they could do that they could literally help change the world."[iii] I drew parallels with this intention, and what happened when I consciously introduced inner awareness to the young people I worked with. I noticed that that it was hard for young people to connect to the reality of climate change without first of all having a sense of themselves and their values. My voice sounded preacherly and the responses were flat and disinterested.

## Shifting perspective

In 2009, Global Generation became finalists in The Big Green Challenge, a NESTA (National Endowment of Science and The Arts) competition with the promise of one million pounds of prize money. The focus of the competition was grassroots community approaches that could show significant carbon savings. We needed to demonstrate tangible results in terms of the amount of carbon emissions that our projects were reducing and also the numbers of local people involved. At the time, we were working in a local all-girls secondary school running an accredited horticulture programme for a group in which most of the girls had been deemed by the school as having special educational needs, mainly for behavioral issues. The girls were not high academic achievers and they struggled in the constraints of the school system. Two months into working with them, my colleagues and I were still struggling to engage them. Each week, we ran a non-academic programme with this group of girls. The outward focus of our sessions was local food production and new green spaces in the city, such as bio-diverse and food-producing living roofs. In most of the sessions, we tried to involve the students in discussion and evaluation of three key aspects:

- The changes in their own behavior.
- Their understanding of climate change.
- The understanding of whatever gardening activity we were involved with at the time.

The reality of actually doing this proved to be very difficult. Despite our best efforts, after two months, the girls didn't appear to have made any real progress. Often my attempts to engage them were met with eyes glazing over and the hollow sound of my own words. They would say things like, "I don't care ... I can't help it I've got anger issues Miss" or "climate change, what's that?" When I posed the question of how we might deal with the fact that our global supplies of drinkable water were at risk, the answer was, "I'll just go to the shop." I realized pretty quickly that I would get nowhere by scaring the girls with the fact that their own existence was under threat. However, there were times when the introduction of a few minutes of silence opened up a positive energy between them, which seemed to carry over into how they worked as a team on the practical tasks. They said things like: "When I stay still and silent, I like it because it is peaceful ... it's like I have discovered a friend inside me."

In preparation for the new term, I spent time thinking about how we might change our approach in the hope that we might find a better way forward. I wanted to encourage the girls to become familiar with the whole of their experience, both the inner and the outer dimensions. I wanted them to see them-selves and what they were doing from a bigger perspective. A perspective which, I believed, would give them the motivation and the ability to change. This approach proved to be a breakthrough for the girls and for myself as a facilitator. In short, it really worked. The girls' level of attention, understanding, and the fullness of their responses surpassed my expectations. Rosemary, our volunteer (who was also a retired teacher), and I felt this was the best session yet. By the end of the session, the girls were animatedly talking about how they could inspire others and how it makes them angry that people don't care.

In the session, I linked the work in the garden and the learning about climate change back to the girls' experience of themselves. They associated their own feelings with the movements of the seasons and the climate, for example uncer-tainty, creativity, and courage. Drawing on the integral theory work of Ken Wilber[iv], who describes three important dimensions of our experience as Self, Culture, and Nature, I came up with the frame of I, We, and the Planet. This helped the girls focus on their inner, outer, and collective experience. Bill Torbert's[v] four territories of experience - Framing, Advocating, Illustrating, and Inquiring – provided guidance for how to structure the session.

**Vision and Purpose** *(Framing)* I asked the girls to define what vision meant to them and they responded by describing how it was giving things meaning and having something you want to achieve. We spoke about personal vision and bigger collective vision. The girls looked at the goals they had set themselves at the beginning of the programme, and wrote their goals for going forward, before feeding them back to the class.

As I wrote their responses on the board, I realised that each of them could be associated with either "I", "We", or the "The Planet."

> I want to fully understand the meaning, for myself, of what we are doing, the vegetable garden and learning about Climate Change.
>
> Joyce, 15 years (I: Self)

> I want to do more teamwork.
>
> Michaela, 15 years (We: Culture)

> I want to create a beautiful vegetable garden.
>
> Joanna, 15 years (The Planet: Nature)

**Explaining** *(Advocacy)* I explained why it is important to have vision and why we need to understand our experience in order to achieve that vision. I also re-visited the different parts of our experience; individual, collective, and the outer tangible world that we can comprehend with our senses (I, We, and the Planet).

**Illustration** I drew three circles on the board, two on the left and one large one on the right, and I got the girls to map in each circle the different types of activities that we had done together:

I – Silence and stillness, learning diaries, distinguishing where they were in terms of levels of choice. In this part of the lesson, I asked the girls to sit in silence and stillness for five minutes.

We – Teamwork, brainstorms, helping each other.

The Planet (what we had physically created) – Designing a garden, making planters, planting vegetables, and installing a water recycling system.

It seemed that I had barely finished the circles before the girls were filling them, enthusiastically coming up with suggestions, realising at the same time how much they had done and learnt.

**Inquiry** In the last quarter of the session, the girls carried out peer-to-peer video interviews about how they had changed since being on the programme, and what they now thought about climate change. They also worked out their carbon footprints and discussed the practical things they were doing to reduce carbon.

One of the girls started telling us about how they were given a hard time by other girls for doing a gardening project:

> They don't realise how interesting this is, there's a lot more to it than they think ... I used to think this gardening thing and the environment was a load of crap, but now I am learning about the whole world, it's really interesting.
>
> Jade, 15 years.

This was the same girl who had said, "Climate change, what's that?"

As the girls framed experience in terms of "I, We, the Planet" they began speaking about how one's sense of "I" can extend to care for plants and animals and so it was a logical step to speak about what we considered to be "We" and in this way the "We" became bigger and bigger. We spoke about what perspective means and what it is to have a small view and only see yourself (ego-centric), to the next level which would be family (tribal-centric), to the UK (nation-centric), to the whole planet (global-centric). I knew in the way the girls were listening that they were understanding and seeing something new. I said to them, "This is why we call our organisation, 'Global Generation'" and an "aha" resounded in the room – in other words, a generation of people who have a global perspective and who generate activities which will support the planet and all who inhabit it. Subsequently, I heard the girls talk about Global Generation as if it were a movement, rather than an organisation.

> If we start doing more, then we will inspire others and they will want to join the global generation.
>
> Joyce, 15 years, 9th January, 2008.

Two months later, the girls hadn't gone back to the level of behavior that caused us so much concern. It was encouraging to get feedback from a visiting workshop leader who I brought in to run a session on communication and public-speaking. Ereni had had many years of experience working with teenagers across the UK.

> At a certain point in a session, there is usually a need to emphasise the responsibility we have to communicate something of importance; the value and power of the individual as a communicator. When I work with a group that has never really thought of this, this is usually wholly new territory and it takes me a while to change the energy and attention in the room. With this group of girls, something extraordinary happened. The very moment I mentioned the power and significance that we all have and the responsibility, I felt the group fall into a palpable energy shift upwards. It was as if they knew what I was referring to and they all immediately CHOSE to go there as a group. It seemed they could access this part of themselves from having practiced it. In this, they had a dignity, maturity, and quiet confidence. I must admit I did not expect this and I was surprised, especially as they were not gifted and talented students but rather SEN (special educational needs) students, so I know this was far more difficult.
>
> Ereni Mendrinos, 4th February, 2008.

During the final session of the year with the girls, I asked them how they felt they were doing. One of them was quite self-aware in the way she summed up the group's behaviour. "Somewhere in the middle of our best and our worst, because we keep going up and down." We didn't win the Big Green Challenge, but thanks to that competition "I, We, and the Planet" became a strong feature of the way we have framed Global Generation's work ever since. Being part of a competition encouraged me to persevere on a project that looked like it was failing. The experience with the girls gave me confidence that it is possible for the perspective of "I don't care" to change to "now I do."

## Going back to the beginning again

In order to go forward, sometimes it is helpful to go back to our roots and beyond, to an imaginative or even experiential sense of the place before anything ever happened. Cultures throughout the ages have held origin stories, which are, as Thomas Berry[vi] says, what gives us the purpose, direction, and psychic energy to find our way into the future. Working on a construction site can bring the future down to a very immediate human scale. I have noticed that our participants, young and old alike, are more able to engage in meaningful discussions about desirable futures for the planet if they have, even to a small degree, experienced a dimension of life that seems out of the stream of time. In 2011, I wrote a journal entry after facilitating a workshop for a group of young people in the

Skip Garden. My aim was to give the participants a sense of this timeless dimension of life. The steps outlined hold true for me many years later.

> *Amid the background sound of trains and cranes, I asked the group to go back … right back to the very beginning, which can be found in the silence inside ourselves. Before the practical part of a workshop began, I asked the group to commit to being really still and completely silent for a few minutes; to relax by letting everything go and at the same time to be totally aware. As everyone committed to this together, a tangible "We" space opened up. I gave simple guidance and encouragement during the exercise; however, the main thing was that I as the facilitator was still in myself. This helped the stillness and silence to be held among the group.*[vii]

15th September, 2011.

David Bohm provides a compelling description of why cultivating an inner sense of stillness evokes an experiential understanding of the place before anything ever happened.

> The entire past is in each one of us in a very subtle way. If you reach deeply into yourself, you are reaching into the very essence of mankind. When you do this, you will be led into the generating depth of consciousness that is common to the whole of mankind and that has the whole of mankind enfolded in it. The individual ability to be sensitive to that becomes the key to the change in mankind. We are all connected. If this could be taught, and if people could understand it, we would have a different consciousness.

David Bohm[viii]

In order to grow a shared "We Space," my colleagues and I have sometimes followed up "sitting still" with a values exercise. Participants are invited to choose a card from a range of cards and to speak about why that value is important to them personally in the context of whatever is being explored in the workshop. Each card represents a particular value. For example, integrity, takings risks, patience. By doing this activity over many years, we have built up a lexicon of over 50 values which could also be called inner qualities. Just as plants grow through the exchange of nutrients, sharing what is important to us can support the growth of community.[ix] Not only is reflecting on values helpful on an individual level, but we have discovered through our workshops (often with participants of different ages and social circumstances) that it is in the sharing of values, that a space opens up where different values connect. This 'values sphere' engenders a hopeful sense of possibility. This is what one of the young people wrote after working on the build of our first skip garden described earlier.

> We created something only a team could do. We created a living atmosphere and a welcoming place. Working with strangers from all sorts of places and backgrounds helped us achieve this, as it brings everybody together with the common cause.

Zak Nur, GG Generator, 16 years

Another useful "We" exercise, is to invite participants to share with each other their experience of the time they have had together. This can be done as a "go round" the circle. It can also be done as a form of dialogue through a no hands up "thread conversation." This encourages deep listening and following the thread of the conversation.

## The three baskets of knowledge

### *My Journal*

*Rod stood in the middle of a field gently guiding a group of 8 and 9-year-olds: "I would like you to stand up very slowly without making a sound ... now you are going to walk in the way that ancient peoples walked. They needed to walk this way when they hunted animals; if the animal heard the sound of a broken twig or a leaf rustling, they would run away. If we walk this way, nature might say something to us. As you walk, I want you to use all of your senses, feel the ground under your feet, the wind on your face ... the sound of the birds ... notice the different grasses and the creatures that live amongst them." In silence, the group spread out; each one finding their own ground. "Now cut your pace in half, go as slow as you can go," said Rod. Coming together again to share their experience, it seemed the world had opened itself to us a little more that day. "I tasted the air" ... "I saw two dragon flies" ... "I noticed the swallows gliding" ... "the grass was as soft as a sofa" ... "the ants moved like lightning" ... "I felt that I was a hover fly" ... "everything works together and has its own special job," said the children.*

*12th July 2014*

Asking the question, "What does the land have to say to me?" would probably have been dismissed as superstitious and animistic by my forebears. This, as Thomas Berry[x] pointed out, is the autism of our times. "I, We, and the Planet" is an invitation to remember that we live in an ecological world and as Martin Shaw writes, at the heart of ecology is mythology[xi]. It is perhaps no surprise then that, in time, I was to come upon a mythic legend that spoke to the reflective practices we had carried out in relationship to "I, We, and the Planet." The traditional Māori legend of the Three Baskets of Knowledge, offers guidance for how human beings might walk in generative ways upon the earth.

Although I have come across different interpretations of the Three Baskets of Knowledge, the common thread is as follows. Tāne, the god of the forest, is asked by Io, the Supreme Being, to journey through the 12 heavens to retrieve the knowledge that will guide human existence on earth. The knowledge he received came in the form of three kete mātauranga – baskets of knowledge – along with two stones for assimilation of knowledge to ensure that what is selected from the baskets is used wisely, not simply for personal gain but for all.[xii]

^xiii*Tāne crossed a silver bridge made from billowy vapour. As the vapour cleared Tāne saw in front of him a beautiful woman. She looked up at Tāne and her eyes shone like greenstone in the sunlight. Tāne thought she must be a messenger from Io and so knelt before her and spoke, "I have come to receive the three baskets of the sacred knowledge. Can you tell me where I can find them?*

*The woman looked at Tāne and her words penetrated deep inside him. "I will show you the Three Baskets, but you must promise to always carry them so that humans will walk well on the face of the earth." And Tāne agreed that he and all of his children and his children's children would be the guardians of the knowledge. "Look into my eyes – what do you see?" said the woman. Tāne looked into her dark shiny eyes. He was shocked; he saw himself staring back at him, all of his hopes, his fears, his attributes, his challenges, and he described this. "That," said the woman, "is the first basket, knowledge of oneself as one truly is."*

*"Look again" said the woman. This time, Tāne saw the woman and he saw the reflection of the sky, the rocks, the plants, the birds, and the trees, and he said, "I see you and all that has passed in all that is present in the physical world around me." "That is the second basket of knowledge", said the woman, "Knowledge and respect of everything around us."*

*"Now I will show you the third basket and this is the most important one. In the future, if people forget this, they will lose their way." Tāne looked again and this time, he saw the woman, the plants, the animals, the sky, the sea, and he saw himself. He described how he saw everything as one thing. The woman smiled and said, "To see one's own reflection as the interconnection upon which the universe is built – this is the third basket, Kete Tuatea."*

Over the years, I have shared the Three Baskets with many different people in a range of settings, from children and their teachers to construction workers. Myth carries a universal license to be told in different ways^xiv, and I have often told the Three Baskets story using images and characters drawn from the concrete terrain of King's Cross. Spurred on by the question of how I could enable others to find ownership of the work, I passed the Three Baskets to others to tell in their own way. The following account is from a camp that Global Generation held for 12 young people at Pertwood Farm. For some, it was a first step into being a Generator (Global Generation eco ambassador). It was also the first time many in the group had been out of London.

On the train, I sat beside Lily, one of our senior Generators. All the way, she wrote. What emerged was her version of the Three Baskets of Knowledge, which would provide an over-arching narrative throughout the camp. This time, the protagonist of the story was a teenager, like Lily. She spent her days plugged into music, drowning out her frustrations with the contradictions of school policies around equality, which were not reflected in the social environment; the corridors were filled with racist comments. Each day, she wished she could lead a life that was worth more. One night, she dreamed of a friendship between a spider and a seagull, and the next morning, the seagull appeared in front of her. Together, they flew out of the city to a gorse and hawthorn copse, with a fire pit and a yurt. As she looked into the seagull's eyes, she saw herself and learnt about the First Basket of Knowledge ... her hopes and her fears, her challenges and the values that would help her find a way forward.

As we sat together that night in the Pertwood yurt, Lily told her story. This was an invitation for the Generators to begin exploring the first basket of knowledge by introducing themselves to each other through the values they had expressed by coming on camp. With their journals in hand, they then considered the question, what does it mean for me to leave London behind? (All of the girls wrote journals during the camp and I have included examples drawn from two of the girls.)

> Camping is very difficult – I don't know if I will be able to sleep at night. I will worry about my family who I am so far away from. There is no turning back; I just have to make the best of it.
>
> Nazifah, 15 years.

> Our tent is filled with bugs and it is cold; the toilet has no light or heater. Many like it, but this has made me realise how much I miss the city, the loudness, the fact that there's always someone creating havoc.
>
> Kulshum, 16 years

The next morning, the clouds hung heavy in the sky. Undeterred, the Generators set off up the track. They made their way slowly to the big stone circle, each taking a place by the stones, standing in silence; not one flinching from the wind and the rain that was now coming down, even if ...

> ... it was too silent; it was nice being on your own for a bit, but I still prefer being in a room full of people or a city.
>
> Kulshum.

> As a city kid, walking over the hills in the rain was quite a journey for me – seeing all the valleys becoming distanced. The different shades of green looked like a small section of a rainbow.
>
> Nazifah

During the day, we explored the Second Basket of Knowledge, the reflection in the seagull's eyes of everything around us, be it rock, person, animal, or tree. In the blackness of the night, we became animals as we left our torches behind and silently threaded through the forest. By the morning, the Generators were ready to stand in the footsteps of the land ... what might it mean to imagine ourselves as the sky, the earth, and all that lies between?

> I am the sky, I am so beautiful in the daytime with white fluffy clouds and at night, I come out with the prettiest stars, but even I have my ugly days, especially when the rain is about. You can always look up to me, whether you are happy or sad, whether you are scared or confident.
>
> Kulshum.

On our last day, a local farmer brought a falcon for the Generators to see. The black shiny eyes of the falcon were a reminder of the Three Baskets. As we approached the final circle, Kulshum said, "Can you tell us the last part of the story – what is in the Third Basket of Knowledge? I want to get into that zone again." This was an opportunity to reflect on all that we had experienced over our time together. As Lily asked the seagull what was in the Third Basket, the seagull transformed into the falcon we had just seen. She looked into the falcon's shiny eyes. She saw the falcon and she saw herself and everything around her. In their own ways, the Generators understood and expressed the Third Basket of Knowledge – the interconnection upon which the universe is built.

> I am the wind that howls and I am the movement of grass that sways side to side. I am what makes and creates things to grow and flourish as I am the ground beneath your feet. I am the soul within you deep. I am the stream in which you fish. I am the world in which you live. I am your home.
>
> Nazifah

> When I first got here, I hated it. I tried to like it but I didn't – the silence, the atmosphere. It wasn't for me … then I thought I am only here for another day. Let me make the most of it. That's when I started to understand what nature is – nature is beautiful. Now I feel a part of nature. I am a city girl with a hint of country girl.
>
> Kulshum

Over the course of the weekend, I observed the power of the Three Baskets to open individuals to a sense of identity that connects us to ourselves, the land and each other as one story. Chellie Spiller describes it like this:

> Across these three orders of reality, the Three Baskets of Knowledge equip humans with the necessary skills and behaviours for living. Importantly, the knowledge in the baskets is of a collective nature and is not solely for individual consumption, but for the greater benefit of society.
>
> Chellie Spiller[xv]

It was very exciting to see how well the Three Baskets of Knowledge story landed. It seemed this was a signature story for Global Generation's work. However, I felt a degree of trepidation working with this particular story. I knew that the legend is part of the corpus of sacred Māori knowledge and as such was traditionally not normally related in public."[xvi] Was I as a Pākehā sharing a story that wasn't mine to share? I was aware that I had not asked permission to use the story. I wrote to Chellie about my concerns and sent her an account of the Pertwood camp. Her response was an important acknowledgement of the ground I was standing in and beginning to share with others.

I love the transformations that have taken place on the journey of your young Generators – I feel very touched by what you are doing. What an honour that the Three Baskets is having this effect. It is a demonstration of how Aotearoa[xvii] is woven into the fabric of who you are and how you are making a difference in the world. You have found your Tūrangawaewae and are now helping others – young and older – to find their place to stand – a place of their own knowing.

Chellie Spiller, 11th November, 2013.

## The triple gem

Over time, I was to discover that "I, We, and the Planet" was a triple gem; found not only in the Māori legend of the Three Baskets of Knowledge but in cultural stories and philosophies from around the world. The ancient Greeks spoke of the integration of beauty, truth, and justice,

> Which can be interpreted in a way that fits the "I, We, and the Planet" approach. Beauty is subjective, interpreted differently by each individual, and thereby corresponding to "I." The pursuit of objective understanding or "truth", can be seen as relating to the outer tangible world, corresponding to "the Planet." Finally, the shared activity of coordinated and just action corresponds to "We." Plato advocated that attention be paid to all three dimensions of experience; from this, he said, goodness flows.

Thomas Berry spoke of three overriding qualities of the universe:[xviii] Differentiation, subjectivity, and communion. As I see it, one of the challenges and great opportunities of community-based regeneration in our cities is the increasing amount of human diversity. Differentiation, for Thomas Berry, is a primordial expression of the universe. He writes, "The universe is coded for ever increasing, non-repeatable, biodiversity as exemplified by the incredible variety of life that has evolved on the earth ... our role as humans must now be to restore the earth's ability to continue its growth towards complexity and differentiation." The second principle of the universe Berry identifies is that of increased "subjectivity". Every reality is not just a collection of objects but a community of subjects. I take this as an invitation to consider the possibility of an interior dimension in the "things" that make up the universe, be that river, animal, or tree. As I have been illustrating through the examples of the young people involved with Global Generation, when we learn to identify with a deeper, less boundaried sense of ourselves is when the "me" naturally merges with the "we." This aligns with Berry's third principal of the universe, which is communion.

## Reflections

As exciting as creating food-growing gardens in unexpected places in the middle of London might be for some of us, this on its own has not been of interest to many of the young people Global Generation works with. We have been able to work with young people in meaningful ways because the work is not just about exterior transformation. One of our young generators puts it like this: "What's different about what we do at Global Generation is, not only do we change on the outside but we change on the inside too" (Ciara, GG Generator, 17 years). Re-framing the inner, outer, and collective dimensions of experience as "I, We, and the Planet" proved to be a straightforward way of ensuring that most if not all of our workshops included some element of inner reflection.

Working with "I, We, and the Planet" as an overarching framework can help in two distinct ways:

1. Bringing awareness to the inner and the outer; the unseen and the seen dimensions of life

   - I - a deeper and more connected sense of identity
   - We - the emerging potential between people and all of life
   - The Planet - leaving a tangible positive legacy in the physical environment

2. Broadening of Perspective

   - I - I only care about myself
   - We - I am part of a wider community that includes not only humans but soils, waters, plants, insects, birds, and animals[xix]
   - The Planet - I recognise that we are all responsible for the planet and I want to take positive action

When a deeper, more connected sense of identity develops, the boundaries around oneself are not so hard and brittle. As a consequence, positive change and feeling part of, or even creating a sense of, community is a more natural thing.

The weekend with the girls at Pertwood was a pivotal step for me, both in terms of my connection to the Three Baskets of Knowledge story but also in terms of my interest in developing the practice of storytelling. Rather than a didactic imposition on the Generators of how they should be, speaking about "I, We, and the Planet" in the form of a mythic story was an invitation to explore. I experienced the way that storytelling can bring us into a "zone" in which there is a felt sense that we are part of the natural world and that nature is our home. This in turn helps us cultivate a spirit of co-operation and contribution.

## Notes

i    Berry, T. (2006). *Evening Thoughts: Reflections on Earth as a Sacred Community.* M.E. Tucker, ed. San Francisco: Sierra Club Books.

ii   Tarnas, R. (2012). *Like a Rolling Stone: The Nobility of the Postmodern-And Why the Scientific Quest Against Anthropocentrism Has Been Pursued with Religious Fervor.* [Online]. PCC Forum. Available at: http://www.youtube.com/watch?v=i1xKwJ9b8No [Accessed: 10 September 2014].

iii  Jaworski, J. (1996). *Synchronicity: The Inner Pather of Leadership.* San Fransisco: Berret Koehler. p.83

iv  Wilber, K. (1996). *A Brief History of Everything.* Boston, MA: Shambhala.

v   Torbert. W. in P. Reason & H. Bradbury (Ed.s) **2001**. Handbook of Action Research. Sage: London. Pp.250-260. The Practice of Action Inquiry. (2001), p.208.]

vi  Berry, T. (1988). *The Dream of the Earth.* San Francisco: Sierra Club Books.

vii  Riddiford, J. (2011). Growing Food – Growing People. In: J. Marshall, G. Coleman and P. Reason, eds. *Leadership for Sustainability. An Action Research Approach.* Sheffield: Greenleaf, pp. 206-214.

viii David Bohm as quoted by Javorski (1996) Ibid

ix  McIntosh, S. (2007). *Integral Consciousness and the Future of Evolution - How the Integral Worldview is Transforming Politics, Culture and Spirituality.* St Paul, MN: Paragon House.

x   Berry, T (1988) Ibid

xi  Shaw, M. (2011). *A Branch from the Lightning Tree: Ecstatic Myth and the Grace of Wildness.* Oregon: White Cloud Press.

xii  Spiller, C. (2011). Tane's journey to retrieve knowledge. In: J. Marques and S. Dhiman, eds. *Stories to Tell Your Students: Transforming toward Organizational Growth.* New York, NY: Palgrave Macmillan, pp. 127-131.

xiii I have re-worked the following 'mythic story' based on a play written by James Barnes (2004, pp.170-176) along with an interpretation of the Three Baskets from Rev. Māori Marsden (2003).

Barnes, J.W. (2004). *Sea Songs.* Westport, CT: Teacher Ideas Press.

Marsden, M. (2003). *The Woven Universe: Selected Writings of Rev. Māori Marsden.* Te Ahukaramu Charles Royal, ed. New Zealand: The Estate of Rev Māori Marsden.

xiv See Armstrong, K. (2009). *A Case for God.* London: Vintage Books. p.3. Armstrong claims that myth carries a universal license to be told in different ways depending on the audience and the purpose. In that spirit I have often shared the Three Baskets using images and characters drawn from the concrete terrain of King's Cross.

xv  Spiller, C. (2011). Tane's journey to retrieve knowledge. In: J. Marques and S. Dhiman, eds. *Stories to Tell Your Students: Transforming toward Organizational Growth.* New York, NY: Palgrave Macmillan, pp. 127-131.

xvi Marsden, M. (2003), p.57 Ibid

xvii Aoteoroa (Land of the Long White Cloud) is the Māori name for New Zealand

xviii Berry, T. (1988). Ibid

xix A definitition of community taken from Aldo Leopold (1949). *A Sand County Almanac and Sketches Here and There.* Oxford: Oxford University Press.

# 5

# A COSMIC STORY

We humans are not lords of all creation but lives woven into the complex interdependencies of a beautiful, unfolding planetary system.

Mary Evelyn Tucker[i]

In 2003, around the time I was setting up Global Generation, I came across the work of Thomas Berry and Brian Swimme, who wrote The Universe Story; a scientific and poetic account of our 13.8 billion-year deep time journey. Their love for our planet and our place in the wider universe resonated with my interest in how to foster meaningful connection to the earth and a feeling for the future. Swimme and Berry described the origins of the universe in a way that brought home the scale of the vast, interconnected story we are all part of, which is still in the making. For me, entering into a story of the universe that seemed to include everything – the inner, the outer, and the collective dimensions of my experience – was thrilling. This was a story that spoke to the silence and the symphonies of life, a story that included joy and pain, a story of uncertainty and immense creativity. My questions were, how could this epic adventure be brought to children and young people? How might this tale of crises and continual change influence the way I approached the challenges and uncertainties I experienced as leader of Global Generation? Thomas Berry[ii] describes how origin stories offer us the psychic energy to find our way into the future. As the accounts in this chapter will illustrate, there is benefit in looking to the rhythms and patterns of nature for guidance in human affairs. That being said, this comes with a caution against inflating any story or philosophy into a grand universal theory of everything. As philosopher Mary Midgley describes,[iii] even Darwin hated "magnificent visions" of this kind.

In the last 100 years or more, we have been faced with an enormous challenge; that of finding home and identity in a reality that is increasingly unknown, emerging, and constantly changing. Approaching the uncertainties of our world with a storied sensibility enables us to stand in a process in which the end result is not yet written. In the best stories, you don't know what the end is. If you knew the end, the enchantment of the story would be over. This chapter offers examples of how people of all ages have found meaning, purpose, and direction in the scientific story of our origins. The accounts that follow introduce ways of bringing to life the enormous, evolving, and often overwhelming story of where we came from. The reason that an origin story of this nature has a place in this book is that I believe this story about diversity, change, and uncertainty has relevance in the context of community-based regeneration.

It can raise alarm bells when non-scientists like myself start introducing mythic sensibilities to scientific stories. Encouragement to continue this work has come from Jonathan Halliwell who is a Professor of Theoretical Physics at Imperial College London. Jonathan describes why he feels a marriage of the mythic and the scientific is important:

> A few years ago, through interactions with Brian Swimme, one of the authors of Journey of the Universe, and Jane Riddiford and colleagues at Global Generation, I became very interested in the Universe Story as a mythological framework which could provide a sense of place, purpose, and inspiration in the education of young people. I believe that science has become so good at stringing together the long sequence of facts which make up the history of the universe that our logical understanding of it has vastly outstripped our ability to truly grasp what it means in human terms. I feel that important steps to fill this gap have been made through the creation of educational events for young people involving the universe story and it has been very rewarding for me to be involved in them.
>
> Jonathan Halliwell, May 2016.

## From small stories to a big story

### *My Journal*

*After several hours, the bus we have taken from the airport stops and we are tipped out beside a jetty in a tiny fishing settlement. The wind is howling and rain looks imminent. Grabbing waterproof gear out of my bag, I clamber into a tightly packed zodiac dinghy with two huge outboard motors. As we bounce up and down on the waves of the wild and choppy sea that lies between mainland Canada and a tiny island in Nova Scotia, I look down at my damp sneakers and have that sinking feeling of being ill-equipped for what is to come. A few days later, we head off to our own solo spots, where we will be alone; completely alone for the next seven days. This is getting frighteningly real. I say goodbye to my friend Kent and as he walks away instinctively, I spread my hands out on a huge granite rock. To my surprise, it seems to speak back to me. I feel the energy of the rock enter my hands. I know that I am not alone. This landscape is harsh, the birds seem disappointingly few and it looks like many of the trees are dying. With little else to focus on, I am drawn into a bigger sense of life, a sense of the earth itself breathing. During the solo, I am accompanied by the ripples on the sea and the gathering of storm clouds overhead. I feel these age-old movements of the earth as my breath and my blood. I crouch in my tent, grateful for my new sleeping bag as the temperatures plummet to five degrees and below. I meditate and I read the one book I have brought with me; "The Journey of the Universe."*[iv]

May 2015

The "nature solo" was a contemporary version of a traditional vision quest which is typically a rite of passage for young males as they enter adulthood in some Native American cultures. Our solo was led by ecologist, vision quest leader, John Milton. It was organised by Kent Williams as part of his PhD research, which focused on the benefits of nature experiences on leaders. At the time, I was faced with a major re-write of my doctoral thesis which had not made the grade for the submission date. After the solo, my attitude changed from one of despair to one of acceptance. I realised it would take as long as it needed to take, and this meant that my final year of writing felt like a blessing rather than a curse.

This was not the first time I had experienced myself as part of something bigger; it has been a repeated refrain throughout my life and has shaped what I bring to my work in the world. Even as a child, I felt the earth was connected to me. I can still remember the mystery of looking up at the night sky and the feeling of the warmth and comfort of the rocks in my primary school playground. These experiences have taken me many years to understand, being sometimes very present and at other times nothing more than a memory. Through my family, I was taught to love the outdoors; we swam in rivers, walked on mountains, and rode horses across wide open spaces. However, the idea that the ripples on the sea and the storm clouds overhead are my breath and my blood was not exactly the creed of the farming family that I grew up in. The land, the rivers, and the sea were often seen through the economic lens of productivity and commodification.

For many years, I was tentative when speaking publicly about the very cosmic story behind Global Generation's work. In the public sphere and to our funders, it was far easier to talk about the social needs of young people and the practical environmental work we do than it was to talk about how, through this, we try to foster a meaningful relationship with nature and the wider cosmos. Inspiration in how to introduce this big and intangible topic came from my husband, Rod Sugden, a primary school teacher, who would come home and regale me with tales of the children in his classroom and their enthusiasm as they realised they were part of a vast, evolving, and expanding universe. Finally, in 2012, an opportunity opened up to introduce a programme of cosmic education into Global Generation's work; we were invited to run a summer project in the Skip Garden for primary and secondary school children. I asked Rod if he would co-lead a Big Bang Summer School with me. Even before the project began, it felt like something different was going to happen. I remember sitting in a plan-ning meeting with tutors and students from The University of the Arts, who were going to help us. In my diary at the time I wrote:

## My Journal

*The context of what we were doing, sitting in a little porta-cabin in the middle of the construction site, grew bigger and bigger. I spoke about stepping beyond specialism to a more integrated and cross-disciplinary way of working. One of the students shared her excitement in recognising the same principles at work on different scales in the universe; from the cosmic to the human. I felt no cynicism in the room. Afterwards Rod described, "Everyone started to contribute to the conversation. There was a fire between us which grew as each person spoke. It felt like the universe wanted us to tell its story. It wanted to be talked about and it wanted as many different people from different walks of life to tell it."*

July 16ᵗʰ, 2012

The Summer School was made up of a mixed aged group of 11 and 12-year-olds who were just about to start secondary school and some older 16 and 17-year-olds who were sixth formers in the same school. In order to bring the scientific story of evolution to life in meaningful ways for the young participants and other King's Cross collaborators, including the construction workers, we introduced different ways of knowing. Workshop sessions combined learning about science, practical activities such as working with plants, wood, fire, and clay, and reflec-tive activities such as creative writing and spoken word, art, and theatre. Details of the twists and turns of the evolving scientific story were shared as a poetic narrative, helping participants to find echoes of their own personal story within the vast movements of the cosmos.

As a way of finding personal stories within the stories of the universe, we used the lens of "I, We, and the Planet." For example, in thinking about the "I" the group discussed their own qualities and values and then discussed the qualities and values that are revealed in the universe. This was followed by freefall writ-ing. Participants were encouraged to write in the first person, imagining them-selves as different stages in the evolution of the universe and focussing on the particular value they had chosen. Through the process of writing, many of the group experienced a shift of identity in which a separate sense of self expanded into a boundaryless and creative process.

When I began, I felt like I had to be open-minded because if I was open-minded, I could allow myself to develop and grow bigger. I also had to have the initiative to expand and grow otherwise nothing would have changed. I needed to be independent and rely only on myself because there was nothing else at the time.

Muslima, 11 years.

As the days progressed other colleagues in the Skip Garden who weren't directly involved in the project started to become curious about what was happening. Our Office Manager at the time wrote me an email summing up her experience.

> Something about the atmosphere in the Skip Garden seems different this week. There is a sense of inspiration in the air. I watch around me and see young students learning about the Universe. I have sat in on some of the sessions, engaged and conversed with the young people, and been moved and enriched by their presence and personal enquiries into the topics being covered. It's true that talking about how the Universe began can bring up contention and opposition, different people holding different beliefs. Bringing this discussion forward at earlier stages, more as an add-on to other projects, it felt quite tentative. But now, in this format it has grown feet. It has become a building block, a more solid platform for many features that we already involve people in, such as creativity, curiosity, and independence. The young people have been showing this in bucket loads. They are engaged, lively, and serious.
>
> Manpreet Dhatt, July 28th, 2012.v

"We are not doing this for you, we are doing this because you are needed to grow a different kind of future" is something I have often said to the young people we work with. The truth of this was borne out in the final showcase for the Summer School, which involved parents and our local business collaborators, including the local construction companies. Afterwards, many of the adults involved commented that they had now learnt how to relate to the Universe Story as their own story.

> Something different happens to the kids here, it's just mind blowing watching them take an incredibly complex topic like the beginning of the universe or understanding DNA, that most of us have difficulty grasping. And here we have 11-year-olds that in a matter of days are able to articulate it and write about it. I can see how they are taking on and feeling the responsibility, to take things forward. Recently I was feeling despondent about bankers but these kids give you hope that the world can change.
>
> Visitor, August 31st, 2012.

## Marrying the mythic with the scientific

Even though I grew up in a household full of stories, ranging from the Greek legends to the Chronicles of Narnia, there was still a large part of me that dismissed stories as childish. However, after the success of our first Big Bang Summer School, I became more curious about the power of story. I started to think about the way early people might have sat around fires, looking up at the stars, creating stories to make sense of why the universe exists.vi This led to an application to the

Heritage Lottery Fund for a project called Stories for a Better World. Engaging young people in cultural creation stories as well as the scientific story of the origins of the Universe was the focus of the project. Whilst it took some persuading for the funders to be convinced that our early origins might be considered heritage, eventually we were successful. In preparation for the project, each member of the Global Generation team wrote a mythic-styled story and we researched creation myths from our own cultural backgrounds. As we sat in small circles sharing our stories, I felt the power story can have in opening up meaning and possibility; evocative enough to take us to new places and fluid enough for us to make it our own.

Stories for a Better World was the beginning of introducing cultural creation stories into our workshops alongside the scientific story of the universe. For example, I have sometimes told the New Zealand Māori Story of how Rangi and Papa, the earth and the sky, needed to separate in order to bring light into the world. As in the photograph below, we often begin our workshops for children with a story followed by a discussion.

Rod Sugden and Jane Riddiford with children from the Space and Nature Club in the Skip Garden– courtesy of Global Generation

The children in our workshops seemed to have no problem in transitioning between what is scientifically true and what might be true in other ways. Whilst science is important, arguably it is not big enough to evoke other important dimensions of our experience; the appetite of children for imaginary stories is testimony to this. Reliance on the literal truth at the expense of other ways of knowing perhaps explains why, as Thomas Berry[vii] claims, the more we have discovered scientifically about the universe, the less it holds meaning for us. Historian, Karen Armstrong, describes how "Myth is about the unknown; it is about that for which initially we have no words."[viii] Storyteller, Martin Shaw, explains how myths contain metaphors which allow us to make the expansive connections. "Metaphor is the great leap, the generous offering of many possibilities contained in one image."[ix] As evocative as the telling of traditional tales can be, there is more to explore. These tales come from a time when we thought the earth was the centre of a fixed and knowable universe. Joseph Campbell famously asked, "What are the myths of our time, the myths of an evolving universe?"[x] When scientific information is shared through stories, it can take on a mythos that re-awakens older ways of knowing. A process of remembering a time when we were intimately connected to the earth. For many years, we held lunch and learning sessions in which children were often teamed up with local employees. Together they explored the life of the garden; the bees, the soil, the seeds. On one occasion, a new wood-fired clay oven was being worked on. While one group group pummeled the cob with their bare feet, another made the starter for sourdough bread. Before we added the water to the mix, I told the story of how it took 9 billion years for hydrogen and oxygen, the two elements of water, to come together. At the end of the session, the group were asked to do some freefall writing with the start line "bread is amazing because." The lunch and learners fell into a focussed silence and all that could be heard was the sound of pencils moving across rough sugar paper. In the sharing afterwards, one member of the group, a quantity surveyor who worked for a large construction firm, moved us all with his memorable line; "Bread is amazing because it started with a love affair between hydrogen and oxygen."

Finding natural ways to introduce this work can often be challenging. Sometimes it is the practical things we are developing in the garden or current events that have provided a doorway for discussing aspects of the Universe Story. One of the ongoing questions has been how to make the work inclusive. This is particularly important in King's Cross where some of the schools have more than 50 different languages spoken, and children belong to many different faiths and cultures. Occasionally their cultural beliefs mean that it is difficult for them to accept the idea of evolution. Many of the children have their own creation stories that are part of their culture or their religion. Introducing the evolution story as one story amongst many different creation stories can help children feel included. The following account illustrates how we have we have actively worked to bridge contemporary and traditional stories.

### My Journal

*It was an extremely hot summers day. Six construction workers and the 15 children on our Summer School were sitting together in the shade of a large gazebo. The construction workers had brought with them several large buckets of clay. This was King's Cross clay that had been dug out of the ground that morning.*

*With her hands in the clay, Julia Rowntree of Clayground Collective told us about the stories clay tells of where we have come from. "The past is embedded within the clay," she said. The week coincided with the landing on Mars of NASA's Rover "Curiosity." It had been deliberately directed to an abundant clay area. Where there is clay, there was once water and, if there was water, clay exploration is most likely to reveal if some form of organic life once existed on Mars.*

*Out of the corner of my eye, I saw Sadim light up. Different parts of his life were coming together. For the first week of the summer school, he had struggled with his curiosity about science and the story he had learned as a young Muslim, that humans were created out of clay. The clay was a link and he leapt upon it.*

*"Isn't it amazing," I said, "the sages in bygone times had no microscopes or telescopes but they intuited that clay was important for life."*

*At the end of the session, Sadim wrote, "I believe in Islam, but this is a scientific story that we can all tell."*

August 5th 2012 (Eight years later Sadim's auntie,
Sharin Sultana came to work for Global Generation)

## Shifting underpinning stories of leadership

As we developed the Universe Story work, my colleagues became more curious about the deeper themes behind Global Generation's approach and wanted an opportunity to go into the ideas for themselves. Every two weeks in our Engaged Philosophy sessions, we took time to discuss these deeper themes, and in the process, we developed our capacity for deeper listening and dialogue. The interest of my colleagues in the underpinning ideas behind the work was important to me as it spoke to the question I held about how I might enable others to share ownership of Global Generation's work. Working with the Universe Story and seeing the effect it was having on children and young people gave me the feeling that we were unlocking a helpful participatory perspective. A perspective in which we all had a part to play in an unfolding creative story. A story made possible through many of the qualities I thought were needed to run a healthy organisation. Contrary to Newtonian ideas of certainty and control, ours is a largely unknowable universe born through inconceivable challenge, unimaginable patience, and the kind of creativity that eventually finds its way through. Above all, it is a connected universe in which the relationship of community is key.

At this time, I was in the early stages of my doctoral inquiry. With gentle-prodding from my supervisors, I was becoming more interested in leadership. Pursuing

a practice-based doctorate which focussed on my work as founder of a growing organisation meant that sooner or later I would need to grapple with the
complex topic of leadership. The Universe Story offered a fresh perspective on
how to manage the uncertainties of organisational life. Scouring the internet for
examples of leaders who were learning from our recent scientific understanding
of the universe, I came across a number of critiques of the Newtonian idea of
determinism and prediction in a fixed and clockwork universe. However, apart
from the work of Margaret Wheatley,[xi] I found very little on the lived experience of leading framed within the evolving story of the universe. As I saw it,
this was not just about scientific ideas; the storied way we were working with
scientific information revealed an interior dimension of the cosmos, uniquely
known in the depth of oneself. It is an objective-subjective reality that includes
and goes beyond the boundaries of individual psychologies. That is, a reality that
has been revealed through observable scientific method, but which needs to be
brought to life and given meaning by each of us in ways that are of relevance
to the circumstances of our own lives. For me, this bigger perspective offered
a different story than expectations of perfection and harmony that I have been
conditioned by. This story made space for the inevitable personal conflicts and
challenges that arise in an organisation. In many ways, I was looking for guidance in human affairs through understanding the rhythms and patterns of nature.
Stephen Toulmin puts it like this; From the beginnings of large-scale human
society, people wondered about the links between cosmos and polis, the order of
Nature and that of Society.[xii]

Relying on the sense of meaning and purpose that can come from having
something bigger to believe in is complicated ground. Like many before me,
I was to discover the potential danger of holding an overarching narrative too
tightly. As I will discuss in the next chapter, singular visions, in the long run,
are never helpful. Drawing on Darwin's autobiography, philosopher Mary
Midgley writes,[xiii] "Darwin understood that large ideas do indeed become dangerous if they are inflated beyond their proper use; dangerous to honesty, to
intelligibility, to all the proper purposes of thought." As I now see it, finding
our roots within the evolving story of the universe is not an idea to be imposed
on others, nor does it live as a name or a theory on the pages of books. It
would be easy to steer clear of such vast and problematic territory. However,
as astronomer Royal and Emeritus Professor of Cosmology and Astrophysics at
Cambridge, Martin Rees wrote; "To be blind to the marvellous vision offered
by Darwinism and modern cosmology which renders the cosmos aware of itself,
is a great loss."[xiv] He goes on to say that we can't leave this knowledge to spe-
cialists, the most important thing is how science is applied, and that concerns
us all. Rees cautions against the common misperception to suppose that there is
something elite about the way scientists think. The opportunity to be inspired
by science and then to use science to think about ourselves in bigger, deeper,
and more connected ways than the scientific narrative offers has been helpful,
for myself and for the some of the young people I have worked with. Here is

just one of many examples of what engaging in this larger, more participatory perspective offers. In describing her 3D art work, which was inspired by learning about the universe in a series of workshops run by Rod, one of our young participants said:

> I have created something which shows how everything is connected, even if the connections are hard to see. It makes me feel that I am not alone because everything around me is connected and we are all made of the same stuff.

> Poppy, 13 years, 2018.

## From straight lines to spirals

### *My Journal*

*Bergson, vitalism, Māori cosmology … I did a double take. My story was becoming coherent. Did the philosophy of a people who didn't consider the land theirs to conquer and own mean they were more often aware of a creative and connected impulse that flows through all things? Was this the same un-knowing knowing that had led me to leave New Zealand all those years ago? I pursued the reference that had caught my attention and a few weeks later, a brown parcel arrived. Inside was a huge red book, "Indigenous Traditions and Ecology – The Interpreting of Cosmology and Community," edited by John Grim. Eagerly I turned to the chapter by Mānuka Hēnare, Tapu, Mana, Mauri, Wairua: A Maori Philosophy of Vitalism and Cosmos. It felt like a major stepping stone in my journey into Māoritanga to come across Mānuka's descriptions of the way in which the "vitalism" (the creative and connecting energy) described so compellingly by Henri Bergson aligns with traditional Māori philosophy.*

*The big red book revealed more connections. I noted with excitement that the foreword was written by John Grim and Mary Evelyn Tucker, John's wife and co-author of The Journey of the Universe.*

*July 2015*

Mānuka's comparison of Bergson's vitalism to Māori philosophy was yet another thread that wove together my life in the UK with my roots thousands of miles away in Aoteoroa New Zealand, roots that until recently were unknown to me. The following year, I met Mānuka and amongst many things he explained to me that in traditional Māori carving, the spiral is never closed, a metaphor I would carry with me, enact and understand more clearly several years later.[xv]

In Global Generation's very practical projects, we often open up a long-time context and encourage contemplation and open discussion about the creative and connective impulse that flows through all things. For example, if we are working with wood, we might discuss how the first forests were created, or if we are working with clay, we might consider how the crust of the earth is the story book of life's great adventures. If you think about what science is now revealing, one could say we are part of an epic adventure that is still in the making. Now more than ever, with the reality of climate change, there is strength in remembering that our origin story is about overcoming crises through creativity and the incredible power of community. That being said, I have been tentative about overtly pushing the universe story work. Sometimes I have left it for months on end. Other times it is the work of others which has helped me find new value in the work.

In the summer of 2018, Rod and I were invited to run a deep time journey session at the Quadrangle Trust. This was part of a weekend workshop for a group of adults, focussed on 'Thinking for the Long Time'. The aim of the weekend was to find a relationship to the future in order to cultivate a sense of care for the generations to come. The participants and contributors over the weekend included a native American elder, physicists and environmental activists and several curators of London art galleries. We began our session with a storied account of our 13.8 billion-year journey. During our presentation, I noticed the quality of attention in the room; everyone was listening attentively. This gave permission for us to share the story in all of its mythic proportions. To embody the empty space before anything happened, we took a long slow walk in silence to the top of a nearby hill. Looking out over the Darent Valley, the group sat in the long grass and engaged in partner dialogue about the qualities they heard in the story. Each person was given an envelope in which they received a small part of the story of the universe, an artistic representation made by two of the young people involved with Global Generation. They dwelt within the part of story they had been given before writing about it as themselves, freefall style. For example, "I am a spiral galaxy ... I am the meteor that destroys." The plan was to then walk the "time line." Instead of walking in a straight-line column, we made our way to the centre of an enormous spiral that Rod had laid out in rope. Bunches of bright yellow ragwort were positioned along the spiral, to mark the points where significant events in the universe had occurred: the original flaring forth, the stars, the sun, early sea creatures up to the present day. We stood together in silence in the empty space in the middle of the spiral before each person found their place at different points around the spiral. One by one, from the original flaring forth to the future, they shared their writing. Each voice, whether it was the voice of bacteria or the voice of a whale, felt like the voice of all of us. The illustration on next page is created by one of my colleagues, which shows how the evolution of the universe can be represented as a spiral rather than a straight line.

Deep time journey – Illustration by Emma Trueman Global Generation Paper Garden Manager

Rituals open up a space for reverence and respect; grounding and bringing big stories alive in our day-to-day lives. We had performed an age-old ritual of reciting our history from the beginning of time, and in doing so we had created a spiral of life between us. A part of each of us was threaded through the story of the universe. The space between us became infused with imaginative and personal meaning that connected the beginning of time to the present, from destruction to new life.

There are no straight lines in the natural world; there are many spirals. Working with the scientific story in a spiralled way helped me experience the connection between all the parts of the story and helped me feel the potency of what Mānuka had explained to me several years before. Traditionally on carved doorways the spirals are not closed. This represents the empty void of Te korekore,[xvi] the empty place before creation that is full of potential. Māori artists represent the cosmic process as a double spiral. Standing together within this potent symbol, we enacted the eternal process of being and becoming.

> The double spiral form is at once an expression of the nature of being and existence, or genealogical connection from the earth to the cosmos and back, and the vehicle by which our world is sung into being.
>
> Makere Stewart-Harawira, 2012.[xvii]

## Reflections

I have come to see that the universe story, at least in the way I work with it, is not a fixed canon of how things are; patterns and structures in the universe resonate differently with different individuals. The Universe Story work, at its best, is an invitation to dig deep, to dare to give voice and imaginative expression to the depths of one's experience whatever that may be. At its worst, it is an abstract and inaccessible set of ideas which have little to do with the embodied and often challenging realities of living on our planet. So, the leadership of the work is a delicate balance of being intentional and letting go, being committed and being flexible.

Sharing an evolution story with children, young people, and adults of different cultural backgrounds requires sensitivity and creativity. In our work, it has mostly, but not always, been possible to find ways of including different viewpoints. Particularly in the early days of this work, I treated science as some kind of absolute truth, which wasn't helpful. Thanks to the input of theoretical physicist Jonathan Halliwell, I have learned that new theories about how the universe began appear each week.

I was particularly heartened by the impact our workshop at the Quadrangle had on one of the young men attending the weekend (an artistic director of an advertising agency). He expressed that he had grown up in a Nigerian religious household and that despite his father being a physicist, neither he nor his family looked to science for any of life's big questions.

> The Deep Time Journey surprised me each step of the way. When you introduced it, I was sceptical. I grew up in a strict Christian household and the Big Bang theory was something that was NEVER discussed! I don't know what it was about your approach, but it felt inviting and made me curious – which is so important … it made me think – WOW! – there is another way.

It has been helpful to find resonance between the scientific origin story and aspects of Indigenous creation stories. Mary Midgley points out that it is often the lay person, like me, who reifies science as true, without further questioning or finding out for ourselves. Working with mythic creation stories alongside the scientific narrative or telling stories of science in mythic ways has enabled children and young people feel more included, particularly those who come from backgrounds where cultural origin stories are valued. A flexible and inclusive approach has also helped them hold their own stories in less rigid ways, sometimes recognising that that specific narratives were often borne out of the understanding and needs of the time.

Mythologist Martin Shaw[xviii] once said to me that traditional myths and associated rituals take thousands of years to develop. Whilst that might be true, as I see it, we don't have that much time and there is a pressing need to develop myths and stories that will re-connect us to the earth and each other. Having said that, my experience is that none of this work is an overnight affair. Looking back on my life, I can see that for more than 30 years, I have asked myself the same questions Joseph Campbell proposed. What are the myths and rituals of our time? What are the rituals through which we might find ourselves within a woven universe?

## Notes

i     Tucker, M.E. and Moore. K.D. (2015). *A Roaring Force from One Unknowable Moment: The story of the universe has the power to change history. Orion,* May-June 2015, pp. 26-31.
ii    Berry, T. (1988). *The Dream of the Earth.* San Francisco: Sierra Club Books.
iii   Midgley, M. (2004). *The Myths We Live By.* Oxford: Routledge
iv    Swimme, B. and Tucker, M.E. (2011). *Journey of the Universe.* New Haven, CT: Yale University Press.
v     Riddiford, J. (Riddiford, J. (2013). *Cabbages and Cranes: Weaving Together People and Possibility.* In: P. Reason and M. Newman, eds. *Stories of the Great Turning.* Bristol: Vala Publishing Cooperative. pp. 159-167.
vi    Swimme, B. and Tucker, M.E. (2011). Ibid. p.37
vii   Berry, T. (2006). *Evening Thoughts: Reflections on Earth as a Sacred Community.* M.E. Tucker, ed. San Francisco: Sierra Club Books.
viii  Armstrong, K. (2005). *A Short History of Myth.* Edinburgh: Canongate. p.3).
ix    Shaw, M. (2011). *A Branch from the Lightning Tree: Ecstatic Myth and the Grace of Wildness.* Oregon: White Cloud Press. p.113)
x     Campbell, J. (1986). *The inner reaches of outer space: metaphor as myth and as religion.* Novato, CA: New World Library.
xi    Wheatley, M. (1999). *Leadership and the New Science. Discovering Order in a Chaotic World.* San Francisco, CA: Berrett-Koehler.
xii   Toulmin, S. (1990). *Cosmopolis.* Chicago, IL: Chicago Press. p.67
xiii  Midgley, M. (2004). Ibid
xiv   Rees, M. (2019). *Beyond The Laboratory* in RSA Journal Issue 1, 2019.
xv    Hēnare, M. (2003). *The changing images of nineteenth century Māori society: from tribes to nation.* PhD Thesis, Wellington: Victoria University.
xvi   Marsden, M. (2003). *The Woven Universe: Selected Writings of Rev. Māori Marsden.* Te Ahukaramu Charles Royal, ed. New Zealand: The Estate of Rev Māori Marsden.
xvii  Stewart-Harawira, M. (2012). *Returning the Sacred: Indigenous Ontologies in Perilous Times.* In: L. Williams, R. Roberts, and A. McIntosh, eds. *Radical Human Ecology.* Farnham: Ashgate, pp. 73-88.

# 6

# ENCOUNTERS WITH THE HIGH PRIESTS

When you really empower people, you don't just empower them to agree with you.

Joanne B. Ciulla[i]

Charismatic leadership, combined with limited self-awareness, is the stuff of cautionary fables. This chapter discusses issues of power related to followership and leadership, empowerment and disempowerment. The stories unfolded over more than a decade, and are part of the reason why I am interested in the dynamics of leading.

Power and boundaries often go together. The first story in this chapter illustrates the need to put up a protective boundary around Global Generation in order to protect what I consider to be a "boundaryless" space. The boundary was strong and invisible and appeared when it was needed in relationship to an incident with the power of organised religion. Considering the need for a boundary raised questions about my own behaviour, prompting reflection about how throughout my life I have in different ways fallen prey to the negative influence of the high priests of power. As described earlier, since I was a child, I have experienced inklings of a deeper ground of knowing, that lay beneath the dominant stories of how things were meant to be. At times, these inner impulses opened a sense of depth, profound connection and a boundaryless sense of freedom and possibility. However, the promise of perfection is tricky territory, as the well-known adage goes; all that glitters is not gold. As a young woman responding to these inklings, I began exploring meditation, Buddhism, and other forms of Eastern spirituality. In 1986, I met and entered into a student association with a spiritual leader, a man I refer to as David; soon I was part of an international community of David's students. In time, the so-called path of spiritual enlightenment digressed into little more than the invidious and all too common dynamic of being a follower in an organisation where cultish behaviour was at play.

This experience raised many questions which I begin to address in this chapter. Questions which I feel have broader relevance in a world in which heroic forms of leadership have held sway. Why did I become a follower of a charismatic leader? Why did I suppress doubts about controlling and oppressive behaviour? One of the ways I made sense of my experience was to write about it in the form of a mythic story. This happened several years after the end of David's community and my engagement with David. Writing about these dynamics in the form of a mythic story was a way of making sense for myself of some of the reasons why followers can easily fall into the trap of handing over their own agency and judgement to other individuals and to rigid ideas born of a single story.

## Boundaries on a boundaryless land

An experience in 2013 opened a consideration about the power structures of organised religion, not just the religion of the 1800s but a modern-day version of the church's colonisation of meaningful spaces the world over:

> (In Santiago) the Catholic cathedral is right where the temple of the sun used to be. That's an example of land-claiming by the Christians. You see, they are transforming the same landscape into their landscape by putting their temple where the other temple was.
>
> Joseph Campbell and Bill Moyers, 1988[ii]

Fast forward and history repeat itself. I was approached by the Bishop's missionary for the new King's Cross development. Father Jack and I had built a connection over the previous months, sharing meals and discussions in the Skip Garden. There was, however, an unsettling question for me – what was his mission? It appeared to be a fairly significant drive, backed by the money and power of the diocese, to have a visible sole faith presence within the development, and the Skip Garden was deemed a suitable entry point for it.

The developers had been wary from the outset, but they hadn't directly said "no" to Father Jack. They had hedged the issue by passing the decision to me. Perhaps, they suggested, the church could go through a pilot period, renting space in the Skip Garden. The church intended to run activities for the residents from the nearby social housing blocks. They had been given the key to enter the secured precincts of the buildings so they could go door-knocking and depositing flyers. Global Generation would provide space, food, and a number of garden-related activities. I initially said "yes" as we had a good relationship with Father Jack and he was prepared to pay a generous commercial rate for the premises.

Early next morning, I woke troubled. How did I feel about having a church banner on our gate? Or to have our name alongside the church's on the flyers that would be dropped in people's doorways? I thought about the young people we work with, many of them Muslim, and I felt uneasy. I arrived at the Skip Garden and almost immediately a conversation opened between my two colleagues, Walter and Silvia, and I. They also had concerns. "Would the residents in the housing blocks, who we are in the early stages of establishing links with, assume we are a church organisation?" asked Walter. "Are we just throwing away the philosophical ground we have fought so hard to create?" added Silvia. I realised that no money could replace the non-physical and often un-named "space" we had created. It is a space where depth is available without signing up to the power structures of religion. It occurred to me that just as New Zealand Māori were baffled at the idea of owning land, viewing it as inconceivable as owning the sky, how can anyone own spirit? As I talked with Silvia and Walter, it became clear that I would have to cancel the rental arrangement.

I immediately phoned Father Jack. He shared with me the internal conversation he had been having with the Bishop and seemed relieved that I was speaking directly about my concerns. I wrote notes as we talked. Father Jack told me how he had been telling the Bishop that the King's Cross development is totally different to any other community the church operates in. He said that the church is in every other community in London but, "Here we have to find a new way, a way in which the church hasn't operated before." "In the past," he said, "we entered a community with muscle, money and a God-given right, and we can't do that here."

As we spoke, Father Jack discussed issues of power and referred to the atrocities in the church's name, which he said were nothing to do with the simplicity advocated by Jesus of Nazareth, who had no organisational power behind him. I explained to Father Jack that this situation resonated with the themes of colonialism that were present in my own background. I asked if he minded if I wrote about our conversation. He said that would be no problem; I have, however, chosen to give him another name.

After the call with Father Jack, I suggested that all three of us, Walter, Silvia and myself, go and write about our experience. We shared our writing with each other at the end of the day. My writing formed the basis of the account above. Walter emailed me what he had written.

> As far as it's important to explore our values together in engaged philosophy sessions with GG, exploring the more "everyday" challenges of life and an organisation together is perhaps more real and impactful for me. Being in a culture where this exploration is done as a matter of course helps and perhaps makes moments and decisions like this more meaningful?
>
> Walter, August 2013.

Input from Walter and Silvia helped me act decisively. Subsequently, our Chair of Trustees, Tony Buckland, said, "I am glad our security system is working." It was a satisfying process as together we unearthed, defined, and stood behind the often invisible boundaries of our work together, boundaries that I feel protect a boundaryless imaginative space. As Silvia wrote at the time:

> I asked myself, what does it mean to be open-minded? And what does it mean to hold on to your principles? We offered our physical space, but what are the implications of offering our space? On a personal level, I was surprised and felt very vulnerable because I hadn't stopped to think about everything and how easily we could let it all go. I thought about the developers; can a charity really become the voice of a big development company to make sure that the formation of a new culture can happen in a new part of London? I felt that this was the strongest instance where, what we have been discussing in our engaged philosophy sessions took place.
>
> Silvia Pedretti, August 2013.

## In the shadow of colonialism

It would have been easy to file Father Jack's story away as a contemporary example of colonisation by the church. However, there was more for me to consider about myself in the story, especially if I related it to fault lines within my own background. I had said to Father Jack, "The colonising dimension of the situation echoed the actions of my forebears." The comment stayed with me; what did this mean for me and how I led Global Generation? Around that time, I came across a well-known quote which is often attributed to the Australian Aboriginal educator and activist, Lilla Watson.

> If you have come here to help me,
> You are wasting your time ...
> But if you have come because
> Your liberation is bound up with mine,
> Then let us work together[iii]

Watson's words challenged me to reflect more deeply on whether my efforts to develop and share an underpinning philosophy for Global Generation's work was no different than the colonialist drive to bring their way of thinking to other peoples. Did I think I had a superior narrative with which I could "help" people? In the early days of Global Generation, many of the ways we worked with young people, particularly in terms of reflecting on themselves, came from my experience. An ongoing and open inquiry was how best to support others, who have taken a different journey, to feel ownership for the work, especially the inner or reflective side of what Global Generation does. It felt important to examine any tendencies I might have to impose my ideas on others. Linda Turiwhai Smith[iv] writes about the need for Indigenous communities to understand how colonisation has affected them:

> The reach of imperialism into "our heads" challenges those who belong to colonized communities to understand how this occurred, partly because we perceive a need to decolonize our minds, to recover ourselves, to claim a space in which to develop a sense of authentic humanity.
>
> Linda Turiwhai Smith, 2012

As the descendant of colonisers, what might it mean for me to decolonise my mind? Smith challenges practices of imperialistic research, in terms of the lone ethnographer in the 1800s and the well-funded efforts of 21st century academics, who have worked on the basis of distanced objectification of Indigenous communities; "Why do they always think by looking at us they will find answers to our problems? Why don't they look at themselves?"[v] Heeding her words, I held a mirror up to myself. Writing from that place allowed an unexpected story to come through; the story of my spiritual search. There are positive aspects of my spiritual journey that have informed the work I do and there are also shadows. The spiritual impulse that revealed a boundaryless sense of freedom also led me into an idealised relationship with a spiritual teacher whose way of operating could be viewed as another form of colonialism. It was a relationship that had negatively coloured my ideas about what it means to lead.

## A deeper knowing

In myths and legends, I identify with themes of confusion and quest, challenge and transformation; themes that are common to cultural stories all around the world. In the Three Baskets Story, Tāne, the tallest tree in the forest, ascends from the earth into the innermost realms of Io's dwelling. The story of Tāne describes the path to become better people. It is also a metaphor for the archetypal journey of the mystic, as he or she travels inwards, seeking ways to find unity with the universe, and to become one with his or her concept or knowing of Io or God or the Supreme Being or the Way.[vi]

As a way of exploring why I had subjected myself to a dogmatic form of leadership, I wrote a mythic styled story about my experience. Mythologising the events with my former teacher, provided language and metaphor with which to surface unconscious forces and associated value systems that shaped this period of my life.

As a young woman, when I felt a deeper and hard-to-name dimension of life, I thought of myself as Vivienne, which I later discovered means "alive."

### The Morepork Story

*One night, Vivienne dreamt of the Morepork calling. She woke with a start and knew what she had to do. Going to the back of her closet, she put on the purple cape given to her by her great Aunt Erica, a spiritual seeker who lived till she was over 100. Suddenly, a fire rose inside Vivienne and she was transported upwards on the wings of an albatross. Flying high above the metropolis, guided by the silvery waters of the River Thames and out into open sea, they travelled till they found the opening of the River Dart. In they went, drawn on by the call of the forest and the gentle undulation of the land, eventually landing beside an old farmhouse. It was there that Vivienne decided to stay.*

*As the green leaves turned to orange and fell in great clumps on the ground, the days shortened, beckoning all inside. The sap sank beneath the soil and Vivienne too drew inward. The days got darker and Vivienne's confidence grew in the vast and mysterious movement beneath the surface. She slowed down, listened, and became a part of it. Her cells spoke to her, telling stories of all they had been. She gathered her cape around her as it was frightening to see herself in the faces of old men and women, rats and foxes, fish, amoebas, and dust. Despite her fear, she stayed still, watching the current of life moving forward.*

*As time went by, she noticed the voices of the people in the farm house; a new wind was beginning to blow and she paid attention. It drew her into the kitchen and she stood silently in the shadows listening and looking. Around a long table, people were talking excitedly, fire coming from their eyes. Outside, heavy rain was falling and thunder boomed in the distance; a flash of lightning hit the window. Vivienne heard a wizard's name and a shock went through her. At the end of the kitchen, a man stood silently stirring a huge cauldron of broth. He turned to her and said, "My name is Ridwar. Tonight, I will take you to see the wizard."*

*In the darkness, Vivienne climbed on to the back of Ridwar's iron grey horse and they ventured forth along the narrow winding lanes. As the branches made patterns above their heads, Vivienne heard the trees whispering to her… tread carefully, listen deeply, and the forest will not forget you.*

*Ridwar, Vivienne, and the grey steed came to a stone cottage. They stepped inside, amidst a crowd of 30 people, squeezed together in a tiny room with a blazing wood fire. After some time, a wizard appeared and sat on a faded green arm chair in the corner. Vivienne didn't understand what he was saying and she didn't particularly like him, at least not in the normal way, but something was attracting her.*

*The next night, she found herself climbing on to the back of Ridwar's horse, making the journey to the tiny room with the blazing fire. On the second night, the wizard spoke to her, "Look inside yourself with a very bright light. Look for the gold that glistens inside the deep dark cave." There was an intense atmosphere in the room: focused, relaxed, and serious at the same time. Something was occurring in the silence.*

*On the third night, the wizard said he would speak with Vivienne on her own; Vivienne told him of her plans to visit the Masters of Tibet and the sages of India. Then she heard herself saying, "Can you be my guide? Can I write to you?"*

*The wizard looked at her and said: "It's already happened, a process has begun, right here, right now, inside your very own self." He showed her a picture of himself and a very old wizard; Vivienne felt a sense of deep familiarity. "I know that face", she said. "It is your own face, my dear," said the wizard. In the distance, Vivienne heard the sound of the owl … morepork … morepork; it reminded her of the poem she had written many years before about the spiritual calling when she was little more than a child. The wizard looked at Vivienne and said, "You have found the pot of gold at the bottom of the rainbow."*

There is more to the story of Vivienne and the wizard which bought confusion and eventually clarity in my approach to the word leadership. Campbell and Moyers claim that a myth is a story about gods and explain that "a god is a personification of a motivating power or a value system that functions in human life and in the universe. The myths are metaphorical of spiritual potentiality in the human being."[vii] In this sense, gods are the personification of intangible forces experienced inside ourselves. This can mean externalising power on to others and diminishing confidence in one's own experience. My Catholic childhood was imbued with the view that there was a god in the sky who knew everything. This god was a he, who had special representatives on earth. As I grew older, I thought I had gone beyond trusting the "high priests of power." When I worked for a community arts organisation, the notion of any form of hierarchy, let alone having a job title, was anathema for many of my colleagues. It was then that I first began asking questions about hierarchy. These questions were also fuelled by a back story of a different order. Despite my seemingly liberal views, a belief in the mythic notion of an omnipotent god dies hard. For me, it was meeting a spiritual guru at a young age that held the promise of paradise. For many years, I was blind to the limitations of the guru, who was an autocratic and very hierarchical leader.

## The King Story

*Vivienne followed the wizard, who became a king and was treated as an all-powerful god. In time, the king gathered large numbers of followers around him. His kingdom was infused with an electrifying sense that, just as billions of years ago in the depths of the ocean individual cells learnt to make the fishes in the sea and you and me, now human beings were coming together to make an even larger organism. The king warned his subjects against becoming too dependent on him and at the same time he did everything he could to make them loyal. The myth grew that his kingdom was free from troubles; it was a paradise in which miracles were taking place all the time.*

*The king liked the idea of being a king just a little bit too much and he made sure no one said a bad word about him. He loved to breathe the fine air of the mountain tops and he demanded that the tower of his castle be built higher and higher till he and his favourite followers could nearly reach the sky. No longer nourished by the rich, dark soil that lay beneath his feet, deaf to the guidance in the rhythms of the land, he grew proud and ever grander in his plans. He sent out royal decrees about his own integrity, no longer caring for the people in his palace; it was all about the kingdom. Meanwhile on the ground, his loyal subjects worked harder and harder to find money to grow the king's tower. The earth became parched and the flowers grew no more.*

*Vivienne never made it to the top of the tower. She wouldn't surrender to the king's will and he banished her from living in his kingdom. In many ways, this set her free. Living in the forest, Vivienne was forced to draw on a more independent side of herself; she found strength in the flourishing of the soil and the seeds. Even so, her devotion to the king held strong and she would always go to listen to him speaking from the golden balcony at the top of his tower.*

*The king spoke out strongly about the lies and selfishness of many other kings who had fallen prey to the corrupting ways of power. Like many before him, he believed he had a plan that would change the world and sent his ambassadors to grow a movement across the far corners of the earth. The problem was that as the weeks and months went by, he never stopped to look at himself and what he was doing and confusion grew all over the kingdom.*

*One day Vivienne heard that a rebellion had happened in the top of the tower. The king's closest subjects noticed that the king wasn't living up to his royal decrees. They banded together and with great courage brought a huge mirror to the top of the tower. The king fought with all his might to look away, he said, "I will never doubt what I am doing." With an almighty blow, he smashed the mirror to the ground. One of his subjects, a quiet brave man, picked up a small shard of the broken mirror and held it up to the king's face. The king caught a glimpse of himself. He saw his own broken face, his hopes, his fears, his pride, and dishonesty. No-one knew if he would ever lead again, but what made people feel safe was that the king agreed to take off his jewelled crown and step down from the tower.*

## Charisma and enchantment

It would be dishonest to deny the influence that David had on my life. For the purposes of this inquiry into leadership and community regeneration, I have deliberately decided to focus on the dynamics of myself as a follower rather than who David was as a leader. That being said, some background maybe helpful. David was an American self-styled Guru, who began teaching in 1986. His approach was inspired by the non-dual teachings of Buddhism and Hinduism and in later years, influenced by Integral Theory[viii] and other contemporary interpretations of Process Philosophy[ix]. His was an experiential path rather than an academic path, the upside being that it was impactful. The downside was that he lacked any form of criticality about his own shortcomings and the limitations of his narrow and certain approach; consequently, David's world was full of contradiction. With the assumption that he had transcended his own ego, much of his teaching focussed on enabling others to glimpse and ideally live from a deeper and more expanded sense of self, beyond the illusion of solidity and separation that that ego can create.

Despite the confusion of a teacher that is not an example of his teaching, there were undoubtedly times of deep fellowship and enquiry within his community. I look back with appreciation on aspects of his message which illuminated my own heart, which at times I experienced as not separate from the hidden heart of the cosmos. It was through David I was first introduced to ideas surrounding evolutionary cosmology which included a meaningful marriage of science and spirit. In time, I reached a point where I knew that if I was to have any real weight, I had to find out for myself what was important to me, beyond the ideas I had inherited from David or anyone else. However, it took me many years to recognise that David's approach was seriously flawed. Rather than focusing on interconnection, David privileged the evolutionary drive towards progress within humanity, at the expense of compassion and care for his students. He had very little connection to the land or an ecological way of being. David tended towards generalisations based on a single narrative and had no qualms about saying his approach represented an absolute truth. Most significantly, he seemed to be unaware of his own behaviour and the confusion and suffering to others it was causing. The only way to survive in David's organisation was to agree with him. At times, I felt uncomfortable and trapped but for longer than I care to remember I didn't admit that. Retrospectively, I realised that David's autocratic style of leading influenced my ambivalence about the word "leadership." Time revealed that his was not an enabling creed. For many years, I laboured under the question, "Who was I to consider myself a leader?" Because of the way David lead, I also wondered if being "a leader" was even a good thing. As described in the "King Story," the limitations of a single-story, guru-centred paradigm eventually caught up with David. In July 2013, after considerable pressure from his closest students, he stepped down from the role of guru and formal leader of his worldwide organisation.

My experience with David made me wary of contributing to cultism, either as a follower or a leader. Cultism is not exclusive to spiritual organisations. Leadership professor, Dennis Tourish, describes how cultism shows up in many organisations. He maintains that the core principle of a cult is a compelling and transcendent vision and he lays out the "interlocking ingredients of cultish dynamics."[x] I recognise all of them from my time in David's world:

- A compelling vision (the vision being of a transcendent or totalistic character, capable of imbuing the individual's relationship to the organisation with a higher purpose).
- Intellectual stimulation (of a kind that seeks to motivate followers to intensify their efforts in support of the vision, compellingly articulated by the group's leaders).
- Promotion of a common culture (a set of norms which specify particular attitudes and forms of behaviour deemed to be appropriate).

Tourish also writes of punitive measures in cults towards non-conformity with the dominant story. These can be subtle and hidden. Why had I ignored some of the behaviour I witnessed in David and those close to him? Jackson and Parry write, "You will put up with leadership behaviour that is at odds with your ideals if belonging to the group is very important to you." [xi]I feel the reason for compromise runs deeper than wishing to belong to a group. The spiritual pull towards an unseen, hard-to-name dimension of life meant I wanted to cross a boundary from the known to the unknown. With David, there were long periods in which I had the intoxicating experience of a more expanded, boundaryless sense of self. To make matters more complicated, leading effectively often includes a sacred quality. Keith

Grint goes so far as to say that the sacred is a necessary component of effective leadership:

> Post heroic models of leadership are unviable if they undermine the sacred nature of leadership and that in turn would destabilize the ability of an organisation to function.
>
> Grint, 2010[xii]

The history of the church and my own experience reveals that the sacred, when associated with charismatic leaders, is inevitably fraught with paradox; it is in one breath empowering and disempowering.

> It is into this permanently unstable world that leaders, especially charismatics, step, offering certainty, identity and absolution from guilt and anxiety to replace – and displace – the moral quagmire and purposeless existence that existentialism reveals. Thus, leaders choose to lead and followers choose to follow and the latter choose to avoid responsibility for leading – though followers may explain their choice as foisted upon them by circumstances, fate, or whatever serves the same purpose. Absoluteness and absolution are the twin promises of this fabled land.
>
> Grint 2010

Of one of my first meetings with David I wrote, "There was an intense atmosphere in the room: focused, relaxed, and serious at the same time. Something was occurring in the silence." For me, the atmosphere in those first meetings (and subsequent ones) with David was enchanting. Leadership scholar, Donna Ladkin claims that "charismatic leaders enchant their followers."[xiii] However, Ladkin cautions that the word "'enchant' can also mean 'to put someone under a spell' or 'to delude' – to dupe, in other words."[xiv] For me, that spell held the promise of connection to a deeper part of myself and an opportunity to belong to a community of like-minded people. Like many religions based on notions of an absolute truth, over time, David's community became a fixed and oppressive regime.

Because of my experience with David, I have often dismissed charismatic leaders as ultimately oppressive and dis-empowering. However, I think there is at times a place for individual charisma, which can give others the confidence needed to cross boundaries into unknown and unchartered territory. The main thing is the type of charisma that is at play. Ladkin describes how charismatic leaders can be either generative or degenerative; the determining factor is whether or not followers are empowered through the encounter or diminished by it.

> Personalised charisma is used for the self-aggrandizement of the leader, whereas socialised charisma leads to beneficial community outcomes … which form is embodied rests with the leader and his or her motives.
>
> Ladkin, 2006[xv]

## Leading in a different way

The day that we had been preparing for had finally come. For more than ten years, both Rod and I had been reading about the Universe Story and now we had an opportunity to host the co-author of the book and the producers of the film, Journey of the Universe. Mary Evelyn Tucker and John Grim were waiting for us outside the hotel we had booked for them. As we walked through the streets of King's Cross, we plunged into deep conversation. "What is it that touches you about Teilhard de Chardin?" Mary Evelyn asked. Inwardly I gulped; how could I answer that? "I don't understand everything he writes but somehow he communicates the music of the spheres," I hear myself say. Slowly we make our way through the grandeur of St. Pancras station, under the huge glass ceiling and up the majestic stairs of the hotel. We cross into the new King's Cross development and I explain Global Generation's footprint from the tops of the buildings to the public spaces on the ground. I feel myself talking fast, wildly weaving from the cosmic to the concrete, from education to enterprise, from the colonialist story of land to the future. I am struck at their quality of attention. They seem to be interested in everything and both Rod and I are finding ways to express more than we often have a chance to say. We had both felt nervous; what would these two Yale professors make of *our* interpretation of the Universe Story? As we sit and plan the film screening and workshop, we would be doing with them two days later, they listen intently and seem keen for us to lead things our way. I take note; they are curious and supportive of different interpretations and applications of the Universe Story. Only at one point does Mary Evelyn intervene: "It's really important for you to show the film as one whole," she says. Rod and I spoke about our experience that evening. We were both surprised and empowered by the way Mary Evelyn and John had encouraged us to lead things in our own way, and there was more to come.

At the end of the Saturday morning workshop in the Skip Garden, we had planned to create the space for Mary Evelyn and John to lead a plenary, which we felt would be a respectful way of summing up our time together. They suggested we sit in a circle. Suddenly, John started to chant, a full-hearted incantation of the Crow people, a Native American tribe that he has a close association with. Next, they asked for Rod and I and two of our young Generators, who had been ambassadors for our work, to sit in the middle of the circle. We sat facing the four directions North, South, East, and West. Turning to the workshop participants, Mary Evelyn said, "I invite all of you to share what you hope for Rod, Jane, Lili, and Nene in their leadership of the Universe Story work." Each person showered us with blessings. For Rod and I, who had spent so many years in the shadow of a teacher who had insisted with the tyranny of an almost military regime that things were done his way, the experience was healing. We were now free to stand in our own shoes. I experienced this simple ceremony as an initiation into our roles as leaders, in our own shoes of the teachings of Thomas Berry.[xvi]

## Reflections

In the incident with the church, I was curious that neither the developers nor the social housing group dealt with the issue directly; they avoided and they deferred. I learned that, even though we were a minor player within the development site, Global Generation had the power to speak up and to say no. Exercising that power meant speaking with colleagues about the uncomfortable niggles I had, holding onto principles and setting down boundaries. This meant I had to back out of an arrangement, as it compromised the spirit of Global Generation's work, even though it would have been financially lucrative,

Experimenting with writing about my own spiritual search in the form of a mythic-styled story helped me understand and articulate things that I might otherwise have held back on. This form of inquiry essentialised the dynamics of leading and following. Depersonalising my experience through the story-making process helped me see the choices I had made as a follower to empower an unhealthy form of leadership. Seeking depth and connection, wanting to change the status quo of fragmentation and separation that I experienced in the world around me, meant I was open to someone who transmitted a hopeful vision of how life could be. As I see it, the upside of the disempowering experience I had as a follower of David is that I learned that imposing one's ideas and dictating to others doesn't work. Consequently, I became interested in what leading collaboratively might actually be. Over time, I realised that it is not straightforward and settled on a middle ground, of encouraging collective leadership, whilst recognising that at times effective action requires an individual stand. There are no perfect solutions and perhaps the most important thing is whether or not we as leaders and followers, are curious and caring enough to engage in critical self-reflection.

The inquiry outlined in this chapter made me reflect on the practices I have introduced to the young people and my colleagues in Global Generation; silent walks, sitting still, storytelling, and freefall writing. Arguably they are enchantments, in that these practices are designed to open a doorway into a less bounded, more connected sense of self. Despite and perhaps because of the experiences outlined in this chapter, I feel that enchantment is called for if we are to spiritually and psychologically find ourselves beneath the concrete conformity and the intense pressures of our world. However, working with these more intangible ways of knowing and involving others in that requires ongoing self-awareness. The power dynamic can easily be one-sided and unhelpful. In this territory, ego thrives and all too easily one can become self-important. I have asked myself what *kind* of leadership is required if I or any other leader is to avoid imposing rigid and oppressive notions of an assumed "right way." This question has made me more attentive to the power I hold as the founder of an organisation and more attentive to the way others interpret this power. If my colleagues were to find ownership and leadership of our work together, I needed to encourage them to develop and work with reflective and other practices in their own ways. Often people like to credit individuals in my position with achievements that are the consequence of many people's efforts. I have tried to give credit to everyone's efforts and I know I have often fallen short. Power is inherent in leading and it is often said that power corrupts. The question is, can we hold that power lightly and consciously and with our eyes open? I take what happened to David and others like him as a call for the humility to let go in the face of the temptation of personalised charisma. I draw inspiration from others like Mary Evelyn Tucker and John Grim who as far as I can see, have kept their attention on the work many people are doing in bringing about a responsible human earth connection, rather than gathering a following around themselves as individuals.

## Notes

i    Ciulla, J. (2014). *The Bogus Empowerment of Followers*. In Ciulla, J. (ed), *Ethics: The Heart of Leadership*. 3rd edition. Westport, CT: Praeger Quorum. p.101.

ii   Campbell, J. and Moyers, B. (1988). *The Power of Myth*. New York: Anchor Books, p.116.

iii  Lilla Watson (Aboriginal activist group, 1970s) The quote was used in a speech given by Watson at the 1985 United Decade for Women Conference in Nairobi and has been often referenced in Global Generation's planning materials, as an approach we would hope to emulte

iv   Smith, L.T. (2012). *Decolonizing methodologies: Research and Indigenous peoples*. Second Edition. New York, NY: Zed Books.

v    Ibid see p.230

vi   Maaori.com http://maaori.com/whakapapa/ngakete3.htm

vii  Campbell and Moyers (1988) Ibid see p.28.

viii Integral Theory, encapsulates the thinking of American philosopher and author Ken Wilber and others who have been influenced by his approach. Steve McIntosh writes; "Integral philosophy is essentially a philosophy of evolution that emphasizes the evolution of consciousness as a central factor in the process of evolution overall. This new perspective is compelling and important because it demonstrates the connection between the personal development of each person's value's and character and the larger development of human history." See: McIntosh, S. (2012). *Evolution's Purpose: An Integral Interpretation of the Scientific Story of Our Origins*, New York, NY: Select Books.

ix   Process philosophy has its roots in the pre-Socratic writings of Heraclites and is most commonly associated with Alfred North Whitehead, who viewed himself as an organicist. Other notable figures influenced by this approach were Henri Bergson, William James and Teilhard de Chardin. It is a worldview in which the process of change, which is referred to by Whitehead as becoming, is more fundamental than the notion of being in a fixed unchanging state. The fundamental reality as William James described it is that everything is an experiential event which is influenced by the past, has a measure of freedom in the present and makes a contribution to the future. See: Bergson, H. (1983). *Creative Evolution*. Translated by A. Mitchell. Lanham, MD: University Press of America. Griffin, D.R. (1998). Process Philosophy. In E. Craig, ed. *Routledge Encyclopaedia of Philosophy*. London: Routledge. pp.711-716.

x    Tourish, D. (2011). *Leadership and Cults*. In: A. Ryman, D. Collinson, K. Grint, B. Jackson, and M. Uhl-Bien, eds., *The Sage Handbook of Leadership*. London: Sage, pp. 215-228.

xi   Jackson, B. and Parry, K. (2011). *A Very Short, Fairly Interesting and Reasonably Cheap Book about Studying Leadership*. London: Sage, p.58.

xii  Grint, K. (2010). *The sacred in leadership: separation, sacrifice and silence*. *Organization Studies*, 31 (1), pp. 89-107 see p.104.

xiii Ladkin, D. (2006). *The Enchantment of the Charismatic Leader: Charisma reconsidered as aesthetic encounter. Leadership,* 2(2), pp. 165-180. p.166).

xiv Soanes, C. and Stevenson, A. eds. (2011). *Oxford Dictionary,* Eleventh Edition 2011, Oxford: Oxford University Press. 1998)

xv Ladkin 2006 (2006) Ibid see p.176.

xvi Since their visit Mary Evelyn and John have continued to be interested in our very urban application of the Universe Story. They have included one of our videos in their programme at Yale and consider Global Generation's work with the Universe Story to be part of Thomas Berry's legacy. They have also introduced Rod and I to a vibrant community of people from around the world who are finding their own ways to explore the magic and the mystery of the scientific story of the universe[xvi]. see Thomas Berry Legacy Projects http://thomasberry.org/life-and-thought/engaged-legacy-projects [Accessed 20.4.20) People we have collaborated with around the world who are working in their own ways with themes inspired by Teilhard de Chardin and Thomas Berry, include Jonathan Halliwell, Jennifer Morgan and the Deep Time Journey and Art Buchbinder's Biology of Story Project

# 7

# LISTENING TO LAND

> It is helpful to recall that there was a time when human beings were much more intimately involved in the landscape, where our very language mimicked and was developed from the music of the earth itself. We not only listen to the earth; it listens to us.
>
> David Abram[i]

My nephew Sean, once asked me, how do you know what to do? His question stayed with me and re-surfaced when I got that feeling sense which is a kind of moral compass. This weighty rudder goes under the skin, carving a different course from inherited expectations of how to be. Independent from external authority, it's a gentle and recurring experience of connection to an invisible and deeply known sense of the whole. And at the same time, there are other winds that move the blood in my veins. I respond to them all and eventually, rain-drenched, find my way home.

Managing conflicting and often inherited impulses inside oneself combined with the expectations of others is complicated ground. In this chapter, I share stories drawn from my childhood and teenage years as a way of illustrating two opposing ways of being that I have experienced in myself in subtle and not-so-subtle ways. One is the opportunist drive of the pioneer, in which land and the species that inhabit it are there for the taking, and the other is a pull that comes from a felt sense of connection with the earth as the living, breathing heart of who I am. The fact is that at different points in my life I have been embedded in both these ways of being. Through the process of creative writing and story-telling, I have been able to explore my lived experience of a culturally inherited mythology about how to treat land, that begged for resolve. In this chapter, I try through words to express feelings that go deeper than words, callings to go this way and not that arising from an older, deeper, lesser-known part of me.

How I understand these conflicting impulses, and where I choose to orient myself in relationship to them is no longer an arbitrary matter. As I see it, the survival of our species, the way we treat others and many forms of life on earth is dependent on the impulses in ourselves that we choose to identify with. There is an art in taking time to let deeper sensibilities surface; feelings that are beyond what the dominant narrative has taught us to value. Through the stories I share, I attempt to show shifts in awareness in which the creed of me as a separate entity opened out to possibilities as we, a collective sensibility, which includes not just humans but soils, waters, plants and animals.[ii] It is about listening to intuitions; hearing a different sound in the silence of the air, feeling gravity in my bones. Henri Bergson[iii] felt that kinship combined with reflective awareness brings about intuition that enables us to realise our true home; a foundation from which we can act positively in the world. I experience intuition as an inner drum beat that sounds beneath more conditioned thoughts.

Philosopher Mānuka Hēnare describes how, "Philosophically, Māori people do not see themselves as separate from nature, humanity and the natural world are direct descendants of Earth Mother."[iv] This remembering of who we are makes it clear why resources of the earth do not belong to humankind; rather, humans belong to Earth. The second half of this chapter is about creating the conditions for young people to listen to what the land had to say to them. Sharing the Māori legend of Tāne's journey to collect the Three Baskets of Knowledge was a way of inviting a group of inner-city teenagers involved with Global Generation to find home and identity in the land. I have included this detailed practice account with the intention of providing guidance and a range of nature connection activities that I hope will be useful for anyone involved in education or youth work.

## Shifting gears

In an attempt to understand history, I evoked through memory, experiences that gave me a sense of adventure in breaking new ground. Looking through an empathetic lens, I imagined how early settlers might have felt. In doing this, I also felt the pain of breaking up the land and also feelings of more restorative rhythms held within land. I began with the question: what was a time in my life that shaped my values? This is the story that came.

Standing on the verandah of our cottage, I looked up and saw a small plane fly by. It was low, as if it was about to land. I felt the urge to follow it and I did, running as fast as I could down the track, past the farm buildings, the shearers' quarters, the old tumbledown woolshed and around the cutting in the steep, craggy hillside. I felt my 8-year-old legs could take me anywhere. Heading down to the river flats, I could see black cattle dotted amongst the tall, pointy carex grass – the heat was beginning to come into the morning and the air was shimmering. There it was ahead of me; the plane had landed on a straight stretch of the track – my heart missed a beat. Maybe I would be allowed to have a flight, I thought as I ran up to the plane. The pilot opened the door and looked down at me. "Can I have a go? Can you take me up with you?" I asked. He shook his head, replying, "Not now, but maybe later you might be allowed to come up with us." He said. The hope that it might happen sped me back to the house. Bursting into the kitchen, I told my mother about what I had seen. She didn't seem to be cross with me for heading off on my own, or for being gone for what seemed like hours. She just looked at me and explained; "Well, darling, that's a top-dressing plane, it's spreading fertiliser, and Dad and Uncle Pip will be going for a flight this afternoon. There is one extra seat available and we need to decide which of the children will go." I was sure it should be me, I had got there first. But my mother just said, "You have to wait and see." The idea of sharing with my brothers and sisters didn't cross my mind. For the rest of the morning and over lunch, I could barely contain myself hoping it would be me. Finally, I heard my father's voice, "Janie" … he looked at me with his thick dark bushy eyebrows and a warm smile, "Would you like to come in the plane with us?"

When I was 8, I did not know that there was another story to tell about the top-dressing plane. New Zealand is a young land that has been inhabited by people for barely a thousand years; the land was pushed by up out of the sea only a few million years ago, bony ribs of rock exposing themselves to the sky. The flesh had barely time to grow before the European settlers brought flocks of sheep to graze the hillsides. They strung wire fences across the rocky escarpments and made enclosures for sheep on land that was arguably never meant to be grazed. I was used to hearing the farmers say,

"How many head per acre are you getting in?" They were talking about sheep and cattle.

"How many tons of super are you laying down?" They were talking about the superphosphate fertiliser, a magical white substance, mined from Nauru, a tiny island in the South Pacific, bird shit island we used to call it. Huge lorries loaded to the gills with super would snake their way along shingle tracks that had been blasted out of the rock by gelignite.[v]

I too had my own way of blasting through the land. When I was a teenager, over one summer holiday, I was lent a big bay horse to ride. I don't remember his name. Before I put on the saddle, he was gentle and sweet, we would breathe into each other's nostrils and I would hope that this time it might be alright. I wore gloves because my hands were blistered from trying to rein him in. Once mounted, it was never a gentle ride; with the slightest encouragement, he would leap forward. Ahead of us was the first fence. At least this and the next one were lowered by spars, and we would be over them before I had time to think. The next field was large; it had the water pump in a small corrugated iron shed and a lone eucalyptus tree. The bay would stretch forward and really start to go. I tried to slow him but it was no good, he just pushed forward and so I went too, trusting that if we stuck together it would be alright. Then we were in the open country, no spars to soften the way, just wire fences. At least there was a nice solid wooden gate. The gate wasn't for opening; it was a safer option for jumping than the wire fence. I heard my heart pounding along with the bay's hooves on the hard ground. The sheep yards would come into view and my stomach would be sick with fear and excitement. I was with the bay now, daring him to go, turning him around in the largest field of all, the old race course, one circle, another circle, did I have the guts to go? I would remind myself: don't think, just focus and stay with the bay. Find our rhythm, find our pace, and let go. We would gallop in long flowing strides towards the sheep yards, which were perfectly spaced to meet our paces – up and over and then up and over and up and over – phew, we did it. Turning around we headed back – up and over, up and over.

I remember one day when we had done this, the big bay and I stood beside the yards sweating and shaking. We both knew we had gone beyond our limits. I wondered if we could go home slowly. I began to notice the long grasses along the side of the oxbow which still had water in it. I saw a pukeko, a swamphen, now not so common, nibbling the grass and felt another rhythm inside myself and the bay felt it too. We ambled, savoring the last rays of the sun that made the grasses go golden. We turned and didn't head home, but onto the main road and across the bridge of the Ruamahanga River; down the slip road and onto the deep gravel bank with the sound of the stones rubbing against each other. The river was low and we paddled at the edge. We headed over to a sandy bank covered in lupins and I dismounted. Sitting down, I held the reins loosely in my hand, and knowing he wouldn't pull away I let the reins go. The bay breathed on my neck; somehow, he knew, we all knew. The sadness rose in me and I started to cry.

The pain runs deep beneath the more visible stories that shaped my life. Currents started to surface. I barely knew my brother Simon; I was 2 when he died. A small and curious boy, playing in a river. I never fully heard the story of when he drowned, but the sadness carried through into all of our lives. Years later, maybe when I was 8 or 9, I cried and cried; my mother let me. It was about Simon was all that I knew. Even as a teenager when I would walk the land where I grew up, I would cry unexpectedly over the story of the land. It hit me in great waves of grief; I was often surprised at how deep it went. Hillsides, stripped of trees, so the earth doesn't stick to the hills anymore; earth running down to the rivers and the seas. Erosion was a word that filled the air. Now the people are getting desperate, waking up to what has happened. In less than 170 years, the land, once an almost deafening orchestra of birdsong, has, in some places, become silent, save for the rumble of agricultural machinery ripping through it. There is also a new and more hopeful turning in this story. Large tracts of native forest are fenced but this time the enclosure is for another purpose: to bring the birds back. Sitting beside my mother in a darkened theatre, the tears well up in me; tears of sadness and of joy. The film is showing volunteers across the country, who have come out to help re-introduce the kiwi. The land is speaking and somehow, we all know.

I sometimes think about these experiences when I walk the land that I was raised on; they bring out an older, more patient, more intuitive side of me. As long as it hasn't been raining, which makes the agricultural chemicals run into the water, I swim in the river. As I feel the water run through me, I know that I am the river.

## The three baskets revisited

Slowing down and learning to listen to what the land has to say is a big part of what I have wanted to share in my work with young people involved with Global Generation. As I have described, many of them are born and bred in London with little or no experience of exploring the natural world. A camp at Pertwood Farm in Wiltshire has often been a step into the unknown; an opportunity to learn how the land can teach us about ourselves, about our relationships with each other, and to see the more-than-human world as a foundation for doing good in the world. As described earlier, the traditional Māori story of the Three Baskets of Knowledge speaks to these intangible inner and relational dimensions of experience and also the more tangible things in the world around us. It is a story I carry lightly, often leaving it alone for long periods of time. In the story, Io, the supreme god, would not tell Tāne, the tallest tree in the forest, the knowledge in the baskets. Tāne had to discover it for himself. I have learned that this traditional legend cannot be imposed, however it's as if the story asks to be told. This happened during the planning process for a camp with a group of girls at Pertwood. Most of them were 11 and 12-year-olds, and they were accompanied by several 16 to 18-year-olds, who had already been through our Generator (youth leadership programme) and were now volunteering their time to support the younger ones. In thinking about the journey, we wanted the young people to experience, my colleague, who I refer to as Shola, introduced an idea which came from Joanna Macy's Council of all Beings[vi]. She wanted the girls to identify with and speak for a part of nature; be it the wind, a rock, animal, or tree. As I contemplated how this could work, I understood the council of all beings within the context of the Three Baskets of Knowledge.

I travelled by train down to the camp, later than the rest of the group. I was on my own and began reading a book by Peter Wohlleben that I had bought some weeks previously: "The Hidden Life of Trees: What They Feel, How They Communicate."[vii] The words grabbed my attention and I became absorbed in tales of the "wood wide web," the vast underground network of roots and mycorrhizal fungi that is vital for the health of the forest. It is now well established that no tree stands alone, that a forest can be viewed as one organism that communicates with the different parts of itself, by smelling, secreting, twisting, and turning in a responsive rhythm of a vast organic order. Many trees in their original habitat live far longer than humans, with roots reaching out far more extensively than many of us realise. Other trees are rendered deaf and dumb having been force-grown in nurseries and uplifted for transplanting in park landscapes and forestry plantations.

Arriving at Pertwood, I saw the faces of the young people peering out from the clearings amongst the Gorse. Soon we were sitting around the fire, where each of us shared why we had come. What quality did we hope to develop while being at Pertwood? Words like silence, patience, and courage were spoken about. Holding in my hand a piece of flint, painted by my mother in 2002 when we held our very first camp at Pertwood, I felt my roots and my responsibility in carrying Global Generation's origin story. On separate pieces of flint, she had written the names of each of the children who came. I shared the word curiosity as I wondered how the girls might draw out and add to our understanding of the Three Baskets of Knowledge. After this activity, we headed down to a place on the farm that we had discovered during the last camp a few months previously. Nestling under a copse of ancient woodland filled with Hawthorne trees, I looked around at the semi-circle of faces in front of us. I began to tell the story of the Three Baskets of Knowledge, introducing it in the following way: this is a very old story, that was given to the tallest tree in the forest before humans came, so that they would walk kindly on the earth.

There was a strong quality of listening in the group and the words came easily to describe what was in each of the baskets. At the end of the telling, I pointed out that the girls had already brought the first basket to life by sharing something about themselves as we sat around the fire earlier in the afternoon. Now we would explore the second basket, the understanding and respect for everything around us. Hally, who had been an intern with us that summer, shared the story of the 24 hours she spent on her own in the woods. During this solo time in nature, the deer came out to greet her and the trees sheltered her. To help the girls listen more closely to the life of the forest, I introduced a tree meditation. Imagining the flow of nutrients, water, and energy within a tree. I described a journey down through the warmth of the humus, into the cold and rocky layer, and on into the molten core of the earth. Feeling the warmth of the sun drawing the sap up and through stretching branches. With a quality of silence established between us, it was now time for the girls to stand up and find their own place within the land. Shola and I invited them to do a "micro solo" which meant spending an hour on their own in the woods, which for some of them was a big deal. Despite their fear of insects and being on their own, their curiosity won over and they seemed keen to give it a go. Bearing in mind the guidance for the Council of All Beings, I described how they could write an ode, a love letter to a creature or an element of nature that might come forward to greet them. Once the hour was up, the quietness of the forest had slowed me down. Letting go of my eagerness to immediately want to know what the girls had written, I decided to give them time to carry their experience before sharing it with others.

As the light faded, the mood in the group became unsettled. Despite moments of togetherness and calm during the day, a cynical attitude still prevailed. How many more times would we hear, "See you in McDonalds," said in jest to the grazing cows. Shola and I were deflated and spoke about how we might change the atmosphere. Might we encourage the girls to work in the way of story? Could we use story to help the girls listen more deeply? In the morning, there had been a sharing of objects, things brought from London that held personal significance. Could the girls remember and add to the story of an object brought by one of their peers? Despite the fireside banter, the group settled down and became more focused; they were ready to explore more. To our surprise, all of the group wanted to do a night walk. I wondered if they would really manage to do this in silence. In a long column, very slowly, we creeped beyond the border of the campsite, into the long woods. I was impressed at the quiet in the group and disappointed that we heard no birds. Where were the owls, the sudden flapping of heavy wings and the darting deer disturbed in the undergrowth? This time the velvety blackness spoke in a different way. I heard the wind in the leaves and it made me think of the pact that Tāne, the guardian of the forest, had made with his brother Tawhiri, the guardian of the wind.

The next morning, we sat around the fire. It seemed natural to ask about the silent walk in the forest the previous night. I wished I could bottle what I heard from many of the girls for a future day. "I felt my body knew the way" … "it was peaceful, it was still" … "I loved being in the dark." Shola then led us into silence as we sat around the morning fire. Eventually, the group became like a circle of rocks and we were ready to go. Walking up the path across a fence and into a huge open field, Shola encouraged the girls to take the time on their own. In silence, we made our way to the stone circle. I spoke of the two stones that were given to Tāne, as a reminder of the three baskets seen in the eyes of Io. I explained how the early people in this part of the world had brought the knowledge in the sky to the earth through the forming of stone circles. Each of us then found a stone and stood in silence. We made our way to the centre and a part of the story that I had never told before came to me.

Now it was time to tell the story of the wind that raged through the forest and broke the branches of the children of Tāne.

> *Tāne bore no grudges. He knew he could not travel alone to the eleventh heaven to receive the baskets; he would need the help of his brother Tawhiri, the guardian of the wind, who had devastated his forests in the winter months. As Tāne expected, his other brother, Whiri, was jealous that he had not been chosen and he mounted an attack. First lightening and then huge hailstones assaulted Tāne. Tawhiri's cool, warm winds turned the hailstones into rain that nourished the earth and Tāne travelled upwards to the entrance of a bridge made of fibrous vabour, beneath billowy clouds coursed like a river across the blue sky. This was the entrance of the eleventh heaven: the place where Tāne must travel on alone. Before the brothers parted, they embraced and made a pact that lasts until this day. As the wind whistles through the leaves of the forest, it will create music to make the hearts of men, women, and children glad.*

Stretching out in pairs, we spoke of what the story said to us and shared our understanding back into the group. Walking slowly back to the campsite across the fields, our seniors, Linda and Akshita, told me about the pressures they felt at school. What they said made me reflect on the paucity of initiation rituals in our society; for many teenagers all they have are exams. As the camp drew towards the end and talk of London entered the conversation, the group became uncooperative with many of the girls reluctant to tidy up. It was hard to imagine how we might close the camp on a positive note. However, Akshita and Linda brought the fresh energy we needed. They started suggesting ideas for how to continue with the journey we had begun and we encouraged them to take a lead. "Read your ode out loud to a partner and find materials in the campsite that represent whatever you have personified in your writing," said Linda. Soon the group were around the fire again, with an array of autumn leaves, sticks, stones, seed heads, and dew-heavy blades of grass. The atmosphere became open and trusting as we guessed and joked about what each assortment of material might represent. Each thing on its own and all things together made sense. We had found many different ways to express the knowledge of the third basket. Each of the girls took their place, and standing with a long staff in their hands they spoke strongly and proudly.

> Dear Pertwood view,
> Thank you for letting me witness such a beautiful view. Through my eyes, I see a stunning horizon and sunset. You bring happiness and joy. You give me something to be happy about. You bring us close together.
>
> Phylisia.

Soon the vans arrived to take the girls to the train. I wasn't there for the final reflective writing on the station platform; this was the last act before the return of their mobile phones. I wondered if the writing might take the form of letters to the girls from the sunset, the trees, the wind, and the leaves. However, as Shola explained to me, this time the writing needed to be from the girls to themselves. As they cycled back into the depths of the first basket of knowledge, they needed to name for themselves how they had experienced themselves.

> Looking back on my Pertwood experience, I showed stillness. On our night walk, I think I showed stillness even though I barely could see anything. I think my value helped me then. Maybe I could use stillness when I am in London and it might help me A LOT!
>
> Ivy, 11

The photographs on next page were all taken by my colleague Silvia. Over the years Silvia has helped many young people discover the healing power of nature both in London and at Pertwood. The photographs show some of the activities which are common to all of Global Generation's camps.

Pertwood campfire with the Generators - Photographer Silvia Pedrettie

Pertwood Freefall writing in the Stone Circle – photographer Silvia Pedretti

Generators walking accross the Pertwood fields - Photographer Silvia Pedretti.

## Reflections

The pioneer and a slower, more gentle rhythm that connects us to the earth and all of life within it run deep. The pioneer is characterised by the excitement of "I am, I can." Another refrain, perhaps in the rocks themselves, is "we are … and we will." In this chapter, I have tried to convey the qualities of being and becoming at play in the underpinning narratives that influence leadership; from the story of a forward driving machine to the more collaborative dynamics of nature. This shift affects the compass through which I make decisions. It affects what I feel, what I pay attention to and what I act upon. In other words, it changes how I know what to do. I think about the challenges I have faced in reconciling these different rhythms in order to grow a collaborative community within Global Generation. There is sometimes tension in finding the balance between the "I am, I can" and "we are, we will"; accompanied by awareness, both of these impulses have their place.

Looking back, I hardly remember the flight in the top-dressing plane. However, my mother's decision to choose me to go up in the plane reinforced in me a "go get it" set of values. The notion that one is rewarded for following sparks of interest has coloured the rest of my life. It was also an individualistic "I am - I can" creed. Sharing and consideration of my brothers and sisters did not slow me in my tracks. I had got there first, so I got what I wanted. As useful as a pioneering spirit has been in making projects happen, it can, as my nephew Sean reminded me, translate into; "If you want to be successful in life you need to be aggressive and take what you want"; a scary message and one that needs to be challenged. The bullish energy of the entrepreneur can override the ability for deeper listening and collaborative working. In the story of riding across the land, I eventually slowed down and came home to a way of knowing and being that lies beneath and beyond the entrapment of the modernist story of separation.[viii] Peter Wohlleben evocatively describes how the communication of trees is based on a much slower order of time; a vast organic time. Pausing to look at trees and listening what the land might have to say slows me down.

It can feel like running against the tide to invite teenagers to embrace nature connection activities, however again and again, I have seen how it slows them down and brings out a more open, creative, and patient energy. I would say that in order to hold supportive spaces for teenagers to experience nature connection, it is important for the facilitator to be grounded in these practices themselves. That is why each year we run a nature retreat for our Global Generation staff, and many of our biannual away-days are spent in the woods. Early morning silent walks have been a gentle way into the silence that runs beneath and despite a busy mind and in that way, it can create space for listening more deeply to both land and each other. William Isaacs who writes extensively about dialogue and silence describes it like this: "To listen well, we must attend both to the words and the silence between the words."[ix] Practices of silence and stillness can connect us to what physicist David Bohm describes as an implicate order which is enfolded within the physical universe; "To stand still is to come into contact with the wholeness that pervades everything, that is already here; it is to touch the aliveness of the universe."[x]

Metaphors of nature are inspiring for me and there is personal work to be done in shifting the narratives that shape who I have become. It is the slower, localised process of un-picking the past as it presents itself in the present. I wrote about the fallout of farming practices in New Zealand because it is a world I am familiar with and have made different choices about. Weaving together threads of childhood experience has helped me understand rather than vilify agricultural practices that seem hard to understand through the eyes of a more recently gained ecological sensibility. Elsewhere I have described the need to survive and establish a life away from the harshness of the industrial revolution that my forebears may have experienced. I underwent none of these hardships. My childhood was comfortable and relatively easy. For me, there are other factors at play that keep an old story going. The rush of excitement in opening new ground is exhilarating. Driving forward, be it on horseback or at the helm of a machine that spreads fertiliser in the name of progress, brings with it an intoxicating sense of power and control that can mask deeper layers of sadness. Now on many different fronts, the planet is speaking out and as I see it, we have to learn to slow down and engage many different ways of knowing, to really hear what the land has to say.

It is the end of the day as I write this, the sun is going down. It is autumn. A sense of mystery is in the air, as all of life draws inwards. Plants relinquish life and send their energies into the ground. Renewed forces will come around again; in this endless cycle of living, dying and living again.

## Notes

i   David Abram, author of *The Spell of the Sensuous*, as quoted by William Issacs Isaacs, W. (1999). *Dialogue: The Art of Thinking Together*. New York, NY: Currency/Doubleday see page 90.

ii  Leopold, A. (1949). *A Sand County Almanac and Sketches Here and There*. Oxford: Oxford University Press.

iii Griffin, D.R. (1993). *Introduction: Constructive Post-Modern Philosophy*. In: D.R. Griffin, J.B. Cobb, Jr, M.P. Ford, P.A.Y. Gunter, and P. Oches, eds. *Founders of Constructive Postmodern Philosophy*. Albany, NY: State University of New York Press. pp. 1-42.

iv  Hēnare, M. (2001) Tapu, Mana, Mauri, Hau, Wairua: Maori philosophy of Vitalism and the Cosmos. In: J. Grim, ed. *Indigenous Traditions and Ecology. The Interbeing of Cosmology and Community*. Cambridge, MA: Harvard University Press. p. 202.

v   Top dressing with superphosphate has led to the eutrophication of rivers, damaging water life

vi  Macy, J. and Johnstone, C. 2012. *Active Hope: How to Face the Mess We're in without Going Crazy*. Novato: New World Library.

vii Wohlleben, P. (2017). *The Hidden Life of Trees: What They Feel and How They Communicate*. London: Harper Collins.

viii See Eisenstein, C. (2013). *The More Beautiful World our Hearts Know is Possible*. Berkley: North Atlantic Books.

ix  Isaacs, W. (1999). Isaacs, W. (1999). *Dialogue: The Art of Thinking Together*. New York, NY: Currency/Doubleday. p.86,

x   Isaacs, W. (1999) Ibid, p102.

# 8

# GROWING A PAPER GARDEN

> Wayfinders seek to recognise the invisible – to reveal what might remain hidden.
>
> Chellie Spiller, Hoturoa Barclay-Kerr, and John Panoho[i]

How did you begin, is a question I am often asked. Developing a new body of work, within a new community and in another part of the city can be a challenging proposition. In 2016, we were invited to explore the possibility of expanding Global Generation's work to another part of London. My colleagues and I used story and storytelling in a far more conscious way than we had done up until this point. This enabled us to unearth and consolidate connections; connections between different people, connections between the past and the present, connections between all of us and the natural world. In often very subtle ways, the immediate personal stories of the children, the older people and the refugees who came to help grow what became known as the "Paper Garden," unlocked hearts and minds and inspired us all to keep going. Storytelling in its various forms provided ways of loosening the realities we were confronted with: a largely abandoned carpark beside a huge, 1980's, metal-clad industrial building, surrounded by communities divided over changes to the area. Traditional stories took us into imaginative territory where older truths lay waiting. Wisdom spoke to us in terms of human experience and also in terms of the restorative rhythms and patterns of nature. Story also provided a way of drawing a possible future into the present. As has been done throughout the ages, we worked with larger than life mythic characters like the Fire Bird and the Yule King, accompanied by seasonal rituals and celebrations that helped embed values and aspirations for what might be.

In many ways, the stories Global Generation worked with in the Paper Garden echo metaphors surrounding the seen and ordered world and the unseen world that is full of magic and mystery. In traditional tales, these polarities, which have shaped our collective psyche, are often framed as the village and the forest.[ii] Storytelling teacher, Roi Gal Or from The International School of Storytelling, taught me to appreciate contrasts and opposites. Opposites are often the seedbed of creativity and seen in the mythic light opposites become necessary components of a worthwhile venture. Roi explained that for good storytelling we need landscapes full of shadow and light, moments that are joyful and others that are sad. I have found that these mythical contrasts echo the rollercoaster of bumps and breakthroughs that have punctuated our real-life efforts to make things happen on the ground. Our Paper Garden story had plenty of the right kind of juxtapositions for a worthwhile adventure to begin. Apart from anything else, to see it all as a story has made living through it a whole lot more enjoyable.

Whilst we were in the early stages of setting up the Paper Garden, Chellie Spiller, ran a Wayfinding Leadership workshop for the Global Generation team, our friends and associates. This age-old way of being, grounding oneself in the natural world, feeling one's way forward, with full-hearted and open-handed presence, resonated deeply with us. We would return to this wisdom practiced by early Pacific explorers, brought to us by Chellie, in the months and years to come. Seeing ourselves as Wayfinders in Canada Water brought meaning, confidence, and patience to a journey in which the end was not yet written; a journey in which we sought to discover "not only the things but also the relationship between the things."[iii] Stories became a way of revealing and recording the hidden landscapes in which we travelled.

Growing the Paper Garden was an ongoing story of facing obstacles and putting roots down in the cracks and crevices that appear in the wall-to-wall, business-as-usual way of doing things. In those spaces, it was human connection that opened the ways through and helped us overcome obstacles. I hope this chapter will encourage a wayfinding spirit and be useful for anyone interested in applying a storied way of being to their practice in community-based regeneration.

## Crossing the river

After many years of working primarily in King's Cross in central London, I was invited to bring Global Generation's work south across the river to schools and community groups in Canada Water, which is on the Rotherhithe peninsula. This almost island-like part of London is formed by a huge bend in the river Thames. An area where the docks used to be and where many of the original docker families still live. It is also an area that is undergoing huge changes. Like many parts of London, large scale regeneration is underway, bringing with it multi-story buildings and an influx of people with a diversity of cultural backgrounds, experiences, and expectations.

The invitation to travel south came from Roger Madelin, with whom I had worked for more than a decade. He is the former CEO of Argent, the King's Cross developers where Global Generation's Skip Garden is based, and now was heading up the Canada Water Master Plan for another developer, British Land. An area of 57 acres will be regenerated into a mixed-use development that will include a new town centre, apparently the first in London for over a hundred years. It is a process full of complexity, opportunity, and enormous challenge. Proposals have been met with both enthusiasm and resistance by the local community. The sheer force of the development changing the face of London is staggering. Like it or not, redevelopment, regeneration or whatever one calls it is happening. The question for me is: how do we make that a positive thing? Designing future-fit buildings is vital but arguably even harder is the more intangible business of growing the spirit of a place. This raises questions about community and inclusivity. How might we begin to close the gap between the us and them that is all too obvious in London's new developments? How might we avoid these developments becoming modern versions of the 18th century enclosures, complete with their own uniformed security forces, as one newspaper suggested[iv] in a scathing article about King's Cross.

In any ecosystem, the most fertile and biodiverse area is the edge, whether that be the waters' edge or the edge of a forest. There is a movement to appreciate the largely unnoticed biological and social life that exists on the edges of our cities.[v] I have been drawn to the edge-lands in the middle of a city; the often out of bounds spaces inside developer's lands. Travelling hopefully, I have chosen to work on the outside of the inside.[vi] This has required a degree of compromise and a vigilance towards recognising lines that I will not cross. In this work, it has been important to form and sustain relationships with local people and the construction and service teams on the ground, as well as the conductors who are orchestrating the development. Roger Madelin is a conductor in the best sense of the word. He has long been a champion for Global Generation's work with children and young people who can often feel unwelcome in these shiny new places. He describes it like this

> Very soon after the skips arrived in King's Cross, I knew that Global Generation were doing something really exciting. Not only were the children and young people seeming to enjoy working in or learning from the gardens but seriously senior locally employed adults were wanting to become involved and physically participate alongside. The interrelationships between the young people, nature, food, dirt, work, and with business people has created something very special.
>
> Roger Madelin

## Myth-making

The invitation from Roger was hearteningly open. It was pretty much, "Come and do what you do and we will help pay to make that happen." How to make "that happen," if indeed we should make "that happen," wasn't so obvious to me. Moving into an area at the instigation of a developer is not always a good idea. This, coupled with the uncertainty of most of the Global Generation team about the value of developing a second home in another part of London meant I had mixed feelings. I recognised in myself the dogged energy of the pioneer and the excitement that brings. At the same time, I took note of the wariness that I saw on the face of my co-conspirator, Nicole. By that time, Nicole van den Eijnde was joint director with me, of Global Generation. Her grounded sense of caution slowed me down, reminding me of the potentially separating nature of individual drive if lingered in for too long. That being said I trusted Roger's patient and visionary approach and felt that his invitation at least deserved some serious consideration.

Listening to the what the land had to say and embracing a storied sensibility helped us make imaginative leaps into what might be possible. Storyteller, Martin Shaw, describes how "myth is not just about awakening a past that is forgotten; it's also about describing the possibilities of the present."[vii] In order to open the ground for these stories to grow, colleagues and I explored the woodlands surrounding the area to be developed and we got to know some of the non-developer people. One of our community chefs, Sadhbh Moore, and I joined "Sound Camp," an overnight event at Stave Hill Ecology Park and with about 100 others, we got up early to hear the dawn chorus. Our staff team and volunteers also spent an afternoon together in the Russia Dock Woodlands writing and telling mythic styled stories of what might be. Stitched together, these stories formed part of a more "formal" proposal to British Land in 2016. Drawing on the best of our experience in King's Cross, we told the stories lightly in the spirit of possibility. None of it is unusual or surprising; what is more surprising is the absence of these kind of opportunities, given the amount of development that is happening in London.

> *Imagine the year 2025. You are in a place that all began with an experiment. The locals were asked to bring a plant that represented a value that mattered to them.*
>
> *A witch brought a tall bright sunflower for she was fed up with other creatures assuming her bad nature. The troll brought chamomile to represent tranquillity and the wizard brought mint to represent fresh ideas and a zeal for life …*
>
> *It is now a place where young people of different abilities are involved in the creation of gardens in the public squares, the top of office buildings and the terraces in front of restaurants. Many of the plants they are using have been tended by local primary school students in an on-site nursery. The planters and the seats are not all the same, they reflect designs created by teenagers.*
>
> *In the spring, bluebells, cowslips, and primroses fill the open spaces, and there is a natural link between the new development and the local ecology parks, woodlands and the nearby city farm. Birds find corridors of juicy berries and insects enjoy the nectar-filled flowers. The gardens are filled with tomatoes and pumpkins, rainbow chard, and black cabbage and a community café cooks with local food surplus.*

*In the autumn, construction and office workers along with local children and young people brew up great cauldrons of jams and chutney. They put on their bee suits and collect vats full of honey. Everyone is invited to a feast. Teenagers of all abilities exchange products in return for money. Others are apprenticed to the people who take care of the plants and the pathways, and the finest chefs in the land. Teenagers and security guards, mothers and grandmothers welcome visitors to the area.*

*Life beneath the concrete starts to grow and by degrees people start to sing a different song. In the reception areas of many of the businesses are large photographs and poetry that tell a story created by young people and office workers of the world they are beginning to see. Gone are the days of consultants flown in from around the world; people in Canada Water are learning from each other.*

*A visitor to the area writes:*

> *Visiting Canada Water opened a doorway to a new way of seeing ourselves, of seeing our business and crucially, its place in the world. It felt like we'd pushed through the back of the wardrobe and found ourselves in a snowy wood. What is happening in Canada Water was the spark that ignited conversations and actions across the organisation. The team knew that experiencing our place in the evolving and connected story of the universe was a watershed moment – that nothing would ever be quite the same as before. Seeing what is happening here between business and the local community has really shifted identity. We know we are in it together – that it's in our interest to see beyond narrow self-interest.*

With some trepidation, and much to the bemusement of some of the Global Generation trustees, I shared our collective story with Roger and others at British Land and we were granted the necessary funds to begin. This included a monthly retainer and money for school and community workshops.

## Wayfinding

As we weren't going to roll the Skip Garden out and impose a pre-determined plan, I began wayfinding. It was an incremental process of meeting local people. I also spent time with the people who would be involved in directly working on the new development, particularly the master planners for the buildings and the public realm: the architects and the landscape architects, the engineers and the planners. In different ways, I told the story of life beyond the back of the wardrobe[viii] and drew out from the people we met other stories of how people might live, work, and learn together.

In this process, I found it helpful to contemplate the attitude of the early pacific explorers; the wayfinders. They made incredible journeys in double-hulled sailing catamarans. Rather than moving forwards along paths laid down in maps, they stayed still and imagined the land coming towards them. With no instruments to measure, they relied on multiple forms of intelligence including intuitive ways of knowing and knowledge transmitted from one generation to the next; navigating by the stars, tuning into the changes in the currents of the sea and reading the signs such as cloud formations and the flight path of birds. In their book, Wayfinding Leadership, Chellie Spiller and her co-authors, describe how this kind of navigation requires us "to set sail beyond the scope of our own knowing."[ix] To that end, during the months of our work in Canada Water, I visited local schools and many great projects, like Surrey Docks Farm, Stave Hill Ecology Park, Russia Docks Woodland, Time and Talents, and the Bosco Centre. Many of the introductions came from British Land's dedicated community executive, Eleanor Wright. It was apparent that there was a network of strong community and ecologically focused organisations in the area. I also hosted many people for lunch in the Skip Garden, I wanted to give them a taste of what might be possible amidst the intense and not always welcoming construction of a new development. As one visitor wrote

> If the goal was to create an alternative to the reality, then you have done it with bells on. The Skip Garden is so peaceful, a true oasis in the middle of a building site. A paradise within a development. I really hope that we can have a working friendship that will plant some of these wonderful seeds over our side of the Thames.
>
> Rotherhithe Resident - 2016.

Our approach paid off. As newcomers to the area, I expected to meet resistance; however, to the contrary, people were encouraging. The idea of finding ways to grow beyond the "us and them" that so often surrounds big business and development was of real interest to many of the people I met with. A vision and an opening were emerging around involving children, young people, and people of all ages in co-creating pockets of gardens and other participatory projects, like a community kiln and a seed saving bank, in the public realm of the new areas of Canada Water that were to be developed by British Land. Rather than the detailed planning of a new development, required by the planning and the funding process, there seemed to be appetite for leaving at least a few things unfinished and open to what might happen.

## Making a paper garden

The first step was to find somewhere to run a programme of school and community workshops. It was December; wet and freezing. Myself and a colleague, Siw Thomas, were shown around the former Printworks of the Daily Mail. As awesome as the press halls were, I found it hard to imagine them being transformed into friendly spaces in which to work with children. We were taken down dark corridors into a dank and musty tool store with a metal roller shutter and no windows. Our guide excitedly asked me what I thought. "Could this be a community workshop space?" he said; my heart sank. However, Siw, who is an artist and a maker, suddenly piped up. Often, she is quiet in these situations, but this time she said, "I think we can do something with this," and we did. Three weeks later, Siw and I were on site with another colleague, architect Dave Eland; we were receiving a load of news print rolls. These were the ends left over after the printing process, with strong cardboard cores and still plenty of paper left on the reel. (Despite asking the Daily Mail, perhaps predictably we ended up securing the rolls of news print from the Guardian.) In honour of where we were and the history of the area, this was to be "The Paper Garden." Thanks to the skill and patience of Siw and Dave, we involved children and young people in making all manner of things out of cardboard and paper – stools, tables, even a woven paper yurt.

Within a month, primary school classes were making their way through the Printworks security gate. We took the children through the press halls and into the bowels of the building, to the now abandoned ink room. They looked up at the labyrinth of pipes and machinery that was needed to pump the ink into two million newspapers per day. We would then make our way into the Paper Garden workshop, the former dank and musty tool store. In just a few weeks, it was beginning to be transformed. As we stood on the threshold and peered in, I sometimes heard gasps of excitement. One of the older children put it like this:

> The Paper Garden was tiny compared to the huge warehouse, but for me it felt magical. It made me feel I could really do something ...
>
> Arthur 12 years.

For the first few months, our work was behind the closed shutter and it was wearing us down. To do our job properly, we needed daylight and soil. Our initial requests to have a space outside met with a tepid response. Finally, I managed to persuade Simeon Aldridge and Simon Tracey, who lease the Printworks off British Land, to give us a thin slip of land at the far edge of the carpark. It had an old, derelict smoking shelter which was burnt and stained and even had a sign saying "no spitting." Soon, the shelter became a polytunnel for growing plants and the beginnings of a garden were established by Emma Trueman, who became the Paper Garden manager. Our little garden was brought to life by classes of children from local schools and older people from the area who we met through a local organisation, Time and Talents. Some of the older people had lived in the area as children and described picking blackberries and raspberries on the site where the Printworks now stands. By degrees, the experience of seeing children coming in and out of the Printworks worked its own magic. One day I arrived at work to find two enormous black metal shipping containers; Simeon had provided the shipping containers and sectioned off a sizeable chunk of the carpark. In an unexpected and informal way, we were given the area to expand our garden. Over the weeks and months that followed, it became a secret garden.

My next request to the Printworks team was for permission to build a wood-burning oven. The answer came back as a definite NO. This time it was the young refugees who came from the Bosco centre that helped change our landlord's mind. On their first visit to the Paper Garden, undeterred, we made stick bread on an open fire in the garden. Cooking the bread on the fire unlocked stories of their grandparents and stories of the places they had come from. One of the refugees told me that he had walked over eight months from Afghanistan to London. I shared those stories with Simon and Simeon and Ed Cree, the Canada Water Asset Manager for British Land. A few days later, I got a message that we had permission to build a wood-burning oven.

The process of enabling a diversity of people and waste materials to come together to make something special has been incremental and necessarily slow, which can be challenging. Designing a genuinely collaborative community process is a delicate and sometimes excruciating balance of how much to define in advance and how much to leave open for what will come out through the workshop process. I would look around at piles of cardboard, a half-woven paper yurt and an unfinished kitchen, and I would worry. I owe a big thanks to my colleagues for not losing their nerve and helping me see the priorities that we should push on with and what we could leave till later. I also really appreciated the patience and understanding of some of our collaborators. They helped give voice to the way we work and why. Alice Botham was the Operations and Community Liaison Manager for the fast-paced music and events venue based in the Printworks. They host thousands of people each weekend in contrast to Global Generation's workshops which are small and intimate. After Alice and her boss Simon Tracey came for lunch in the Skip Garden, it was encouraging to hear her description of what she saw and how she translated this into what Global Generation had been doing in Canada Water.

> *Going to the Skip Garden made me understand the cement between the bricks. I understood the way that Global Generation creates and builds community through ethos and relationships with different sets of people like developers, children and young people, local councils and local volunteers. It also made me see what can come out of being very slow and allowing people to create at their own pace in an organic environment. What you are doing in the Printworks, I get it now. I get all the little stools in circles … build, the tables … build. The huts … build and everything else … the twig pencils, the scrap paper notebooks all made by everyone. It makes total sense. Seeing the Skip Garden and the way that it works, it's a living breathing thing with a café and a garden and people coming in and out and it moves. I get why the work needs to be slow. It gives it substance. Branches can fall off and things can go wrong but the work has got real roots so it can still live.*

Throughout the process of growing the Paper Garden we have consciously layered in different ways of knowing. The three photographs on next page are a visual record of some of these ways of knowing; crafting, storytelling and creative writing.

Nompsy Chigaru weaving the Paper Garden 2017 – photographer John Sturrock

Storytelling in the nearly finished Paper Yurt –Photographer Lee Carter

Kathryn Oluyinka and Sarah Boukhemma GG generators freefall writing in the Paper Garden yurt – Photographer Tania Han

## Story and ritual

These threads of experience wove together as stories that have shaped what we are helping to open up in the area. Stories that carry values and possibilities which are echoed in traditional tales. In order to help the children settle into the Paper Garden and to open an imaginative space between them, we have developed a regular practice of storytelling at the beginning of our workshops. A class of 30 just fits inside our woven paper yurt. As the children sit down, the atmosphere changes and a sense of quiet falls into the space. The children often comment that they feel like they are inside a nest and in many ways they are. It is a nest of new beginnings borne of nourishing ancient understandings. To honour the soil, the stars, and the seasons we have imbued our very human stories with myths and legends; stories which help us awaken magic in the garden and bring meaning to the changing of the light.

In December 2018, on the eve of the Winter Solstice, I entered the Paper Garden workshop and was transported into a shimmering oak forest … each leaf, each section of bark had a story to tell. My colleagues Emma, Cathy, Joss, and Rod had walked their way into the shortest day of the year by working with hundreds of children to grow a life-sized Paper Forest. As darkness fell, their Mums and Dads, brothers and sisters passed through the Printworks security barrier and were led to a little wooden gate. Unless you were shown it, you wouldn't even know the gate was there. Once inside, they entered our garden, our real garden, where the wood-burning oven and three massive open fires blazed. This was now the winter forest where the yew tree is the queen. She is the guardian of the secrets of the forest; the everlasting tree that is the bridge between the sky and the earth. As fate would have it, there is a yew hedge just inside that little wooden gate. Once the families had made wreaths out of the branches from this evergreen forest, the children stood up tall and proud. They shared their poems in honour of the yew and the passing of the Yule King's crown from the yew to the oak. The adults stood captivated by the children's words as they spoke of the wisdom and the beauty of the trees. Even though, we couldn't see them, I think, on that occasion, we all felt the presence of our ancestors, the twinkling of the stars above the forest. As one of our guests wrote afterwards; "It was great to see such a mix of people; enjoying this local community gestalt/wholeness/glue … I am not sure what to call it."

Our seasonal celebrations have become annual features in the changing area of Canada Water. They are imbued with the spirit that comes from combining children's imaginations and the intelligence of nature. As shown in the three photographs below they have been inspired by mythical characters, like the firebird and the original green father Christmas. We appreciated the fact that the landscape architects, engineers and architects who worked on the Canada Water Master plan joined us for these events. Their attendance gave me confidence that together we were shaping the special role children and young people would play in bringing heart and soul to the public spaces in the development.

Paper garden by night 2017 – Photographer Brendon Bell

The fire bird in the Winter Solstice Procession, made by Siw Thomas, Cathy Wren and children in Paper Garden Winter Solstice 2017 – Photographer Brendon

The Yule king of the 2018 Paper Garden Winter Solstice aka Peter Mitchell – Photographer Brendan Bell

## Reflections

From our work in King's Cross, we knew what it can feel like to have people parachuting in and dreaming up projects. So, in developing the Paper Garden we didn't roll out a formula based on what we had done in King's Cross. We knew the work had to be locally specific; a response to the people and the place. This was made possible through a funding arrangement that we suggested to the developer in the form of a monthly retainer, which enabled us to meet many local people and get a feel for the area before we decided to commit to beginning a new initiative in the area. Going forward, funding for the work has come from different budgets including the community pot and more strategic, long-term public realm funding as well as grants from trusts, foundations, and local government. Achieving a mixed economy, with funding from different sources, enabled us to be a critical friend rather than beholden to the developer.

It gave credibility and confidence for people to see what Global Generation had already created in the Skip Garden; however, at all times, we needed to be mindful that even though Canada Water is still in London, it is a completely different part of the city, with different needs. Whilst we shared principals in how we as Global Generation worked, in terms of our "I, We, and the Planet" approach and daily practices like cooking and eating together, we were clear that we were not offering a formula that could be rolled out. This also meant that we were not going to create another Skip Garden, although in time, two of the original Skips did end up in Canada Water. It was helpful to be explicit about the fact that we wanted to build on what would emerge amongst our participants in the workshop process. As well as helping us step beyond the familiar it helped manage our own and other's expectations. This was particularly important in the first year, when our workshop space looked like a building site.

Another big lesson was in having the courage to be different. By 2016, we were a very different organisation than we were when we began the Skip Garden in 2009. This new step of expanding to a different part of London gave us an opportunity to step beyond what we as an organisation thought ourselves to be. New opportunities lay in the fact that along with being gardeners, chefs, and educators, a few people in the team were creative makers and budding story-tellers. Storytelling, making and related seasonal celebrations were to become defining features of our community-building work in the Paper Garden.

In terms of generating ideas and creativity, it was helpful to draw upon old traditional stories and scientific stories of our origins. In the Paper Garden's very practical projects, different types of stories have helped us paint a big, long time context. For example, in learning about the passing of the mantle from the yew to the oak in preparation for the Winter Solstice celebration, we discussed with the children the rituals and celebrations of the people who had lived in England hundreds of years ago. We also discussed how the first forests evolved on earth. In this sense, the creation of the Paper Garden seemed like a tiny microcosm of what is now being revealed about the creation of the universe. A story that is about overcoming challenge through creativity and the power of community.

In writing this chapter, I revisited my original proposal to British Land. This made me realise that the mythic story of the people in the public spaces, written by the Global Generation team on our away-day some years earlier had become imprinted in my imagination. Not as a fixed plan but as a north star of navigation, helping me know which winds to set our sail to.

## Notes

i Spiller, C. Barclay-Kerr, H. Panoho, J. (2015). *Wayfinding Leadership: Ground Braking Wisdom for Developing Leaders.* Auckland: Huia

ii Harrison, R, P. (1992). *Forests: The Shadow of Civilisation.* London: The University of Chicago Press

iii Spiller, C. Barclay Kerr, H. Ponoho, J. (2015) Ibid

iv Garrett, B. (2017). These squares are our squares: be angry about the privatisation of public space. In The Guardian Newspaper https://www.theguardian.com/cities/2017/jul/25/squares-angry-privatisation-public-space [Accessed: 21.4.2020]

v Farley, P. and Roberts, M. (2011). *Edgelands: Journey's into England's True Wilderness.* London: Jonathan Cape.

vi De Certeay, M. (2011). *The Practice of Everyday Life.* Berkeley: The University of California Press.

vii Shaw, M. (2011). *A Branch from the Lightning Tree: Ecstatic Myth and the Grace of Wildness.* Oregon: White Cloud Press.

viii Lewis, C.S. (1950). *The Lion, the Witch and the Wardrobe.* London: Harper Collins

ix Spiller, C. Barclay-Kerr, H. Panoho, J. (2015). Ibid

# 9

# IN THE JAWS OF THE CORPORATE DRAGON

When we withdraw the earthy metaphors that need to be wrapped around us like a cloak, the thin air of the literal feeds us many untruths.

Martin Shaw[i]

A big part of thinking about community-based regeneration in our cities is the question of what it takes to grow a sense of community amongst a diversity of people in spaces that can feel like they were designed to keep some people out. There is no simple formula: the work calls for localised and nuanced perspectives that go beyond the default position of us and them, above all it calls for patience, persistence, and sometimes precociousness. This chapter is about a project Global Generation is working on with a property developer in West Euston, a changing area of Central London. By the time we were introduced to this 11-acre commercial campus, it was 30 years old, occupied by finance, media, and tech companies and run by an estates team who amongst other things had a big responsibility for security and health and safety. Unlike our work in Canada Water, which involved creating community projects to underpin the implementation of an area of new development, this time we were being asked to help transform an already established commercial campus.

When I first visited, I felt like I was stepping into the jaws of the corporate dragon. For colleagues and I, it was hard to imagine bringing the wider community into what felt like a very alien place. Despite enthusiasm on the part of the developer and generous financial remuneration, we had no idea of how or even if we should begin. Gradually we found ways to be ourselves and to weave a storied sensibility amidst what initially seemed like an impenetrable proposition and a gentle form of activism began to grow. The specifics of what we would actually do were not clear in advance but looking back I could see a kind of natural logic running through the way the project developed. As in the re-wilding of abandoned areas of concrete, we found cracks and crevices where we could begin the work; places that were official but still under the parapet. Rather than working in the main plaza, we chose to operate in the lesser-known spaces. Sometimes we were disheartened as the wild and spontaneous was almost extinguished amidst a deluge of restrictions in the guise of health and safety requirements. However, time and again we were encouraged to keep going by the enthusiasm and very real practical support that came from some of the people who were responsible for the running of the campus, including a large construction team that was working on-site.

Before going any further, I want to flag up that this chapter could read as if Global Generation was working in barren soil where no other community engagement was happening, which was certainly not the case. Perhaps because of the very corporate architecture of the place, the extent of wide-ranging community relationships and opportunities that were already established by the time we arrived on the scene was not immediately obvious. Much of Global Generation's work was made possible through the groundwork that had been done and through the generous support of the on-site community manager who worked for the developer. Having come from a community background herself, she was a connecter and an innovator and she understood the challenges we were up against and helped us find ways through.

Some of the accounts in this chapter have been written by my Global Generation colleagues. They were the people that weathered storms and brought magic to grey places. As described earlier, it is a conscious part of our approach to gather stories as we go and to support not only young people, but also our staff and volunteers to engage in freefall writing as a way of capturing experience as it happens. There are many stories that could be told; I have chosen to share ones that speak to a process of being invited in, finding it difficult to move and through hanging in and the behind the scenes footwork of many different people, trust grew and things changed.

## Feeling our way forward

In early 2018, I stood with members of the Global Generation team at the foot of concrete towers in a shiny granite land, two miles west of the Skip Garden. This was Regent's Place, a commercial campus that sits on the edge of the busy, noisy Euston road; a home to large international businesses like Facebook. Although there were no signs to say keep out, to me and my colleagues it felt unwelcoming; however, appearances can be beguiling. Knowing that things needed to change, Juliette Morgan, who headed up Regent's Place for British Land, had invited us into to the jaws of this corporate dragon. Fortunately for us, it has turned out to be a friendly dragon. I felt that Juliette, who was relatively new to the post, had a heart and a passion to find ways for Regent's Place to genuinely benefit the community, and she was committed to making a positive stand for the environment. I liked the fact that she wasn't precious about the place and above all, I wanted to back the boldness and bravery of her vision about what might be possible. I was to discover that her vision and a very real concern for the condition of the local community and our planet, was shared by many of Juliette's colleagues.

On that first day, as my colleagues and I stood in the concrete and the rain, I noticed my attention shift. From looking for clues in the sharp angular lines of the granite blocks that surrounded us, my attention turned to how alive and engaged the conversation was between us; everyone was saying what they really thought. We didn't all agree about how to approach this new step, or even if we should move in this direction. It was clear that this was about far more than coming up with a simple design proposal for a meanwhile community garden in the main plaza. Each contribution was considered in terms of integrity; not what would be the easy thing to do, or the financially lucrative thing to do, but what would be the right thing to do. I noticed the energy between us, together and different, and I began to let go and listen more deeply. I realised I didn't have to be the expert leading the charge, but rather I needed to be committed to creating spaces for honest conversation to grow and that the right response would emerge between us. Looking at the lone security guard reminded me of my hope that at some point he would be involved in the cooking, the growing, and the conversations with young people. The air was filled with glimpsed potentials and very real concerns.

> It feels wrong to import the community into this very alien space. It's a space designed to keep people out – we can't force them in. ... Could we support the animation of a "green line" that would run from Regent's Park through to King's Cross and beyond?

The challenge felt both exciting and daunting. I felt something good could happen but I didn't know what. Even though we had been through this process several times in other places, it was still hard to imagine how to begin. As I wrote about the experience 12 months on, many of our original questions had been answered:

> Maybe the start should be a kitchen – where we can feed people and begin discussions? ... Where we could sit in circles? ... Where could the stories begin? ... Could we get support for a group of Generators for Regent's Place?

Despite our reservations about what might be possible, I came to an agreement with British Land that in the spirit of action research, I would spend the first couple of months getting to know people and Global Generation would do some low-key projects in the wider community. Much of this was made possible through the many introductions given to us by Regent's Place community manager, Rose Alexander. We made a plan that I would send Juliette a proposal about how we might take things forward. Contracting was deliberately loose to allow enough space for the work to evolve, which for me was a clear signal that British Land would be a committed, brave, and exploratory organisation to collaborate with.

## Growing a different kind of story

I did write a proposal, with lots of nice pictures and quotes from people endorsing the work, however something seemed lacking. Rather than promises of perfection, I knew our job was to rough up the angular precision of the looming towers; opening spaces in which the raw reality of the world as it is might come through. In line with eco-philosopher Freya Mathews,[ii] I consider this to be the territory of enchantment. By considering a large and connected picture, an underlying sense of enchantment came into view which held new and hopeful meaning. I felt the wild twin Tatterhood,[iii] who hails from Norse mythology and for me carries the earthy spirit of the troll,[iv] breathing in my ear, emboldening me to step beyond polite convention, daring me to break my own boundaries in order to open the ground.

> When the Queen finally gave birth, it was to twins. The first twin to be born was Tatterhood; she was red and wild. She arrived in the world riding a goat, with a wooden spoon in her hand, shouting "meat, meat, meat!" Everyone was horrified and so was I. Tatterhood was not nice, she was a rogue and a rule-breaker. Over time, the King and Queen and Tatterhood's golden sister came to love her as they understood the role she had to play.

In an article entitled "Why Leaders Need Not Be Moral Saints," Terry Price[v] writes of rough, edgy, and wicked qualities that we sometimes admire in people. He goes on to describe how leadership is about creation, risk, and change and is seldom careful and conservative. As I pondered what to do, I walked across Hampstead Heath and ended up in the allotment my husband and I share with a wonderful gardener who happens to be a spoon carver. Amidst the kale and the forget-me-knots, the salsify and the marsh marigolds, a naked sense of enchantment worked its way within me and a story came. It was about a wise witch.

> *Once upon a time, not so long ago, in a land inhabited by mothers and children, cats and dogs, foxes and mice, a big change took place. With a rumble and a roar, down streets narrow and broad, the ogres came. They were determined, once and for all, to drown out the old stories, the whispering secrets that were carried in the wild plants and the seasonal changes of the land. They brought with them big blocks of shiny granite and boiled up cauldrons of concrete and spread it all around. They built huge glass and steel towers that blocked out the light and made cold winds blow fiercely. By day, people came to the towers to make up new stories and issue decrees that sent ripples across the land. These were the kind of stories that would make people all the same.*

*It was a scary place to be. Even though there were no signs to say keep out, the local people stayed away. The mothers and their children started walking around the shiny granite land. Even the cats and dogs were scared. They were afraid that, if they went inside, their hearts would be turned to stone.*

*Months and years passed and the people of the towers grew grey. They knew something was wrong, but they didn't know what. No matter how hard they tried, their new stories didn't make them feel very well at all – the magic potions evaporated and a great sadness hung across the land. The people were getting restless. Fortunately, there were a few brave souls sitting in the towers who still remembered the wisdom of old. They prayed for the return of the Wise Witch – the one who knew the ways of stone and concrete and had the power to release the old stories that had been buried in that land.*

*She answered their call and came. She waited and she watched, with one foot on the concrete and another planted deep within the belly of the earth. She guided the blazing passion in her heart with the cooling waters of time. As her mind settled, she spoke words true and strong. She called upon the trolls, the elves, and the pixies. They set up camp in the open spaces around that cold and granite land. They worked with the children who would build and protect the fires, in the hope that one day they might break through the concrete … which by now had seeped into the hearts of the people in the towers. The trolls, the pixies, and the children ate together, grew plants together and they feasted upon the stories of the sky and the soil. They sang and they danced, and some days they ventured in with wagons to the cold and granite land and shared their potions and their plants … and they brought the fires with them, too.*

*The people in the towers got curious. Each day they looked longingly to see if the magic would come. Some days, the Wise Witch and her helpers persuaded them to stop work and come outside; eventually they began to play. A dream grew amongst them that one day the wagons, the potions and the plants, the trolls and the pixies and most of all, the children might stay. They started coming down to the ground to listen to the bees, to cook together, to grow feverfew and bergamot, amaranth and sorrell – herbs that soothed their aching hearts and brought magic to their minds. The people of the towers threw away their charts and their scales. They learnt from the creatures of the sky and the soil how to work in creative and more connected ways. Together with the cooks and the cleaners, the security guards and the taxi drivers, they began to hatch their own plans to find the life within the hearts of the very same ogres that had made that grey and granite land. Wild and enticing stories brewed between them, their message savvy and strong. These were the stories that were told around the campfires on the nights that followed days of love, labour, and song. No longer looking down at their feet, now the people, young and old, often looked up at the stars. As the blazing lights of the city dimmed, they began to dream and to wonder about the stories cherished inside each other and the larger mysteries of the universe.*

Somewhat tentatively, I emailed Juliette the story and the next day sent her our proposal. The following week we met up. As we discussed the designs in the new black and white buildings on either side of the walk from Granary Square to the

Skip Garden, Juliette suddenly said, "Thank you for the story, it really landed with me and aligns with the direction I would like us to take. I am also happy to take forward the things you have suggested in your proposal, which we will need to discuss in more detail." She agreed to pay a monthly retainer to Global Generation, for work I described as "weaving and pollinating" and additional budget as required for the design and delivery of specific projects which would help grow the roots over the next few years of a community-oriented public realm.

## Finding lesser known spaces

From the get-go, there was an outline plan that Global Generation would involve children and young people in creating some of the gardens in public spaces. This was partly driven by the association that we already had with the Landscape Architects who were leading on the master planning for the overall transformation of the public realm. The reality of Global Generation's involvement was some way off. The final design and planning permission for the public realm was at least a year away which was good for us, as most importantly we needed time. It would take time to grow a locally based team, time to feel our way into the local area and build relationships. Whilst still in Camden, West Euston is not King's Cross. We also needed time to get to know the business occupiers and the estates team who worked at Regent's Place. Very quickly it became apparent that we needed to find the less manicured spaces, the cracks, and crevices in which to begin our work, and we needed to be ready for the opportunities that would come our way.

My appreciation of the natural world and the wider universe has been influenced by the writings of Catholic priest, Thomas Berry,[vi] who liked to describe himself as a "geologan," which I take to mean a spiritual custodian of the earth. Berry writes of how the Arcadian view of paradise was the foundation for significant counter-cultural, ecological movements such as the flower people of the 60s and many who left the city and returned to the land to pursue organic farming, healthy lifestyles and the potentials of a solar age. In my view, this can foster a fellowship which rules some things in and other things out. There's no doubt that through leaving the city behind, young people involved with Global Generation have had transformative experiences. However, that is only half the picture. An interest in growth and change drew me to the dust, noise and down-to-earthness of construction sites in the heart of the city. Some of the people who work on these sites know they are in an industry that needs to change. At Regent's Place, an opportunity to be in the midst of the gritty world of construction came sooner than I had anticipated.

One Friday afternoon, a call came from Robert Townshend, the principal of the Landscape Architecture firm doing the Master Planning for the re-development of the public realm at Regent's Place. Robert asked whether I would be free the following Tuesday for a meeting as he wanted me to meet with a few people including Alec Smith from M3 who were the Project Managers. The project managers handle the flow of money and the delivery of planning and construction; in short, they are useful people to know. Robert was concerned that Alec might not understand the way Global Generation proposed to work. Alec was being painted as the straight guy, the money guy. This seemed to be a way of saying, "Jane show them what Global Generation can do and don't scare the horses." Even though PowerPoint is not my favourite medium, I managed to cobble together a few slides and find a projector in preparation for the meeting, which was to happen in the Skip Garden. It was mid-summer and the day of the meeting was bright and sunny; the Skip Garden was looking good. Alec arrived along with Paul Jaffe, senior asset manager at Regent's Place. As often happens, most of the important conversation took place as we talked our way around the garden. By the time we were inside, there was no need for persuasion and little need for the slides; there was an appetite from both Alec and Paul for doing things differently and what we were hoping to achieve was easily understood. As everyone was leaving, I could see that Paul was visibly moved by our conversation and Alec was trying to persuade me: might we come and visit a gantry on the massive redevelopment of 1 Triton? For the next couple of years, this enormous building in the heart of Regent's Place would be a construction site and Alec had the idea that Global Generation might create a garden with and for the construction workers and with and for the local community. This was an opportunity for children and young people to enter the forbidden land that lay behind the hoarding. In most cases, decisions about where to work comes down to the feeling between the people involved, and I was pretty much convinced before we even went on site. What we found on site was hard to resist; a huge scaffold gantry, one story up with mature trees from the pavement beneath rising up through the scaffold boards. Most importantly, the project director from construction company, Lendlease; Chris Carragher, was bubbling over with an infectious brand of can-do optimism. In a heartbeat, Julie, GG's head of gardens at the time, and I said yes. However, the reality of making this happen took months filled by days of serious doubt. What on earth were we trying to do in an environment where time and safety are the primary concerns? If we didn't know that we had our backs covered by Chris, Alec, and Paul, we might easily have given up.

### My Journal

*I am sitting here writing this in our little kitchen, a space we have inhabited at Regent's Place in what was a former Nat West Bank Unit. I look around and see shelves with olive oil and spices, all the cooking accoutrements to help make this place feel like home. I have just cooked lunch for the team on site. This week things are looking up, the first phase of the Gantry Garden is up, running, and working. A welcome relief – two weeks ago I was sitting here with a very sad team who were sodden with yet another downpour of rain. "They will never allow children up there … I don't think we will be doing phase two of the garden," they said.*

26th February 2019

I had been there enough times to feel the discomfort of my colleagues; it called on a deeper engagement from me. It felt like a delicate balance of listening to the concerns of the delivery team who were working against the odds, whilst not losing sight of the reasons why we had chosen to work on this project in the first place. Not for the first time, I took heart in the words of Teilhard de Chardin, whose life's work was dedicated to understanding the challenge and the beauty of the evolutionary process. "Blessed be you, you harsh matter, barren soil, stubborn rock: you who yield only to violence, you who we are forced to work if we would eat."[vii] Thinking about the process of creating fertile soil on barren ground gave me faith that we might just find a way through and be all the better for it.

Our carpenter, Glen McDonald, and Denise Chester, our production gardener, give a flavour of what working behind the construction hoarding was like.

> Glen – It was always going to be challenging, planter boxes on a scaffold platform, not just scaffold but a canter-levered gantry. An inspiring idea and not so easy to bring to life. We didn't get off to a flying start, a few obstacles had to be overcome and the weather was against us too. "Why do you want planter boxes on a scaffold? What's the point?" we kept getting asked. I was beginning to wonder myself as the rain dripped off my hard hat, collected along my collar and ran down the back of my neck.
>
> Denise – It was cold, raining hard and the winds were storm-force. All I could see around me were office blocks and high-rise residential buildings. No wildlife, no colour, nothing green – it was a long way from our usual working environment and it was a difficult place to work, let alone build a garden from scratch. We faced loads of challenges – moving all the plants and equipment up onto the site; clearing security and gaining access; building, lifting, and moving heavy planters which triggered severe back pain; and being called "darling," "sweetheart" and even "the flower ladies" on a daily basis were just a few. We worked extremely hard in very difficult conditions and when we were finished, we felt broken.
>
> Glen – It wasn't all bad though. Alec from M3, Julian and Emma from Lendlease cast spells and wove some magic, the scaffolders from Benchmark went above and beyond supplying us with all the boards we would need. The client site office was a place to warm ourselves; the tea and coffee and chat was much appreciated. "The planters you've made look great!" one of the scaffolders said sliding another plank on to a pile of boards. "I'm gonna make some like that ... Well, I'll have no trouble getting the boards," he laughed. We talked through the basic idea of the planters, the best screws to use and how to build one. A trickle of people coming from the site offices continued to admire and ask about the plants and planter's boxes. John from the client office was particularly encouraging, photographing the planter boxes and explaining he could get his wife to make a few. There is a request from those in the client office ... if there are any plants left over, they would like to have some in small pots on their desks. The flowers and planter boxes will become a fantastic wildlife habitat drawing in bees, insects, and birds. Green spaces on scaffold a waste of time? I don't think so and I believe we have a few converts.

*Denise – Having returned weekly since the end of the build to care for the garden, I've been able to see the end results start to bloom, and I can enjoy some of the feedback from people working within the site. The fruit trees are covered in gorgeous blossom, the clematis flowers are beautiful. The violas add a burst of vibrant colour and the lavender and rosemary plants perfume the air in a way that is totally unexpected on a construction site. The whole of the walkway looks amazing and adds something really special to an otherwise grey, concrete environment.*

*We've had lots of great comments and feedback – from "it looks so good", and "it really makes a difference to my working day" and "they smell amazing!" to the most recent conversation I had. Initially we were met with scepticism by some of the people working in the offices onsite. They were pretty unfriendly and appeared fairly dismissive of who we were and what we were doing. I was chatting with one of these people on my last visit and he said he'd heard there was going to be a phase 2, he said he was really looking forward to it and couldn't wait to see it!*

*Seeing how nature and plants can break down barriers between people like this, and how it smashes stereotypes – watching a six-foot-something huge male construction worker bend down and gently smell the herbs is an image I'll remember for a while.*

After the first phase of the build, I visited the Gantry Garden with Juliette. Walking along the scaffold amidst the planter boxes, our pace slowed. She stopped and rubbed the herbs between her fingers, remarking, "It feels good." She went on to say that she had decided to write a book. "I think I will call it The Concrete Witch," she said. A highlight for me came two months later, receiving an email from Mark Mortimer from M3. His words confirmed my belief in the spirit Tatterhood. This time she had lived in the hearts and minds of many and together we really had broken a barrier. "Alec and I were up on the gantry on Monday and I took the attached picture because I couldn't believe I was seeing dappled sunlight, flowers, and plants on a construction site. Such a contrast from so many other sites we see. Equally impressive is the effect it has on people – sometimes we have some quite heated meetings on site but it's hard to arrive or leave the meeting in a bad mood when you walk through the gantry now."

Though we had worked on the King's Cross Estate for ten years, there were significant differences to working at Regent's Place and we had a lot of catching up to do. In King's Cross, Global Generation's work had largely been in our own compound, a discreet fenced-off area, where within reason we could pretty much do as we liked. On the Gantry Garden, the work was embedded within a construction site and we needed to adhere to very stringent rules of operation. Many of them there for good reason, as construction is a fast-paced and dangerous industry. As newcomers, we had no choice but to put the team through a range of costly trainings to enable them to meet the standards required and to have CSCS cards which are the ticket to being allowed onto construction sites. Whether it was building on the gantry or bringing children into the other spaces we inhabited at Regent's Place, the initial response from those who do the daily tasks on the ground has understandably been one of caution. Often the response to our inconvenient and unpredictable requests came in the guise of health and safety demands, a catch-all for resisting activities that are out of the ordinary. Countless emails have come our way, all needing to be dealt with in a level-headed way. Alongside steamy mutterings, I managed to say to myself, "They are only doing their job."

> *May I just ask who has agreed for this activity to go ahead within this area? in terms of cooking I will need permits to be in place with risk assessments. You will also not be able to carry out the art activity as this will obstruct the fire exit which is a health and safety hazard.*
>
> *Just for your information in order for activities of such nature to take place we must be made aware within a good amount of time (at the very least 48 hours) to ensure that the health and safety aspect of the works have been covered sufficiently.*
>
> *Although a cooking activity has taken place previously on the Campus, we will still require updated RAMS and permits in place for any activity that goes ahead each and every time that the activity takes place. We would also require either a list of names of those attending or at least the number of people attending and their age ranges so that we are aware, should there be any reason for an emergency evacuation.*

Again, we would have easily given up if we didn't know Geoff Jones, the Estates Director for Regent's Place was looking out for us. On several occasions, when our plans were turning to custard and it looked like we had nowhere to operate from, I called Geoff and in an unbelievably short amount of time, wheels turned and doors opened.

I am not alone in my belief that there is a balance to be found between freedom and control. Despite the layers of procedures associated with education and construction, a tacit agreement amongst many of our collaborators that rules and regulations are not necessarily the doorway to behaving safely and ethically is noticeable. A construction manager once said to me: "Our industry is saturated with procedures. We have to find a new way to get people to change, to care, whether it is so they can be more safe on site or sustainability." I have found that as laborious as they can be writing policies, procedures, risk assessments and method statements creates a healthy pause in which to consider how best to proceed. Nonetheless, the question for me is how to grow a generative culture that encourages an independent interest to care for others and the environment, in ways that are appropriate to each situation as opposed to imposing a uniform and potentially deadening set of rules. In this regard, Josie Gregory writes: "Codes of practice often remove the requirement for individuals to develop and exercise their moral reasoning, so they remain morally immature." [viii]This is echoed by Monica Lee, "I am concerned that codes of ethics do not necessarily promote ethical behaviour, and they do not always provide the individual with appropriate guidance on the ethical thing to do when confronted with a real ethical situation."[ix]

## Finding Freedom

In spite of the huge efforts that are made to tame the messy and unpredictable, the wild creeps beneath the surface, and excluded members of the community find their way through the cracks in the concrete. This reveals how the enchanted is incapable of being eliminated and, according to philosopher Patrick Curry, illustrates the failure of the modern mind-set as a constructed reality.[x] In many regards, the aspiration behind the request for Global Generation to work at Regents Place was the hope that we might create a space for enchantment. "I would like you to be the custodians of the spirit," Juliette said in one of our early meetings. Whilst an honour to be considered in this way, it is a tall order to deliver on something which is perhaps best left un-claimed and un-named. Curry writes of enchantment as that which gives life its deepest meaning, is most moving, inspiring, and renewing, and is what one feels has intrinsic value, in and for itself, independent of how useful it might be for some other purpose.[xi] He unequivocally describes a condition/world that is radically non-cartesian[xii] in that it is both material and spiritual, embedded and embodied, and he also states that it is fundamentally "not safe." He argues that enchantment cannot be understood by the calculating and mechanistic moves of the modern mind because it is un-biddable and un-tameable, and as such does not fit with the modernist project of certainty and control; where modernism is present, enchantment is not.

With regard to Curry's view that there is no possibility for recognition of enchantment within modernism, I am a little more hopeful. The Indian sage Krishnamurti wrote, "You can't invite the wind in but you must leave the window open."[xiii] If the conditions are right, it is possible to let go, at least temporarily, of the limitations of what it might mean to be modern. Creating the right conditions is about building trust over time. When any of us let down our guard and our role identities, older, more intuitive ways of knowing are available. This is enchantment as I understand it, and is what I feel my colleague Silvia Pedretti has done in the way she has nurtured early beginnings for our work at Regent's Place. Silvia is our youth and community co-ordinator, responsible for facilitating weekly sessions with the Generators; young green ambassadors aged between 9 and 25. She also runs a range of community workshops, deals with many of the health and safety emails cited above, sets up photography exhibitions and temporary campsites in the bottom of buildings, drives a converted milk float, hosts guests and many more things in between. This is how she describes her experience of finding a spirit of freedom at Regent's Place.

*And suddenly, with the quiet daintiness of a little nocturnal beetle finding its home in the middle of the jungle, or a nostalgic childhood memory of old ideas moving to the surface, a milk float from the 1970s appeared at the end of the road I was standing in, blowing away the last pieces of straw that had been left on the pavement.*

*Behind me, Global Generation's Urban Campsite sat still at the bottom of a shining building in Regent's Place, Euston. The two rounded headlights felt like true beacons of enthusiasm.*

*Regent's Place? A milk float? A bell tent? Straw bales? Are you going to take that on? Are you sure? I still remember some of my colleagues' faces. I had been offered the opportunity to work in a new part of London, with the promise that there lived a wise witch, the head of campus, someone who believed in Global Generation.*

*After a few meetings, the Urban Campsite was installed. And a few minutes after the last paving slab was placed, a group of young people arrived from the local estates, guided by my new colleague Mariam. They were straight in the milk float seats, trying out the lights, the horn, the indicators. Then, straight into the tent. I explained what the space was for, and what we were hoping to do in it with them, knowing that its success depended on their contributions and creativity, and that the reward is rarely instant.*

*Now, Regent's Place is my new place. New people, new projects, new processes, new enquiries. But the core values stay the same. Last week I cooked for a group of visitors, and that really felt like going back to the core of what we do, where the origins have always helped me to find direction for the future. The wayfinder's ethos has helped me to understand this – but instead of a canoe, I will be driving that milk float (shown in the four photographs on next page). It will be navigating around Euston in search of new islands of sanity. I am expecting big waves. Risk assessments, permits for deliveries, funding to be approved, more young people to recruit, and all the excitement won't make for calm seas.*

*So far, we have hosted primary schools for design workshops about what the milk float might be, involved some very local young people in thinking about social action events, had many, many meetings, and finally, have started sanding down the milk float, so it can be painted and refurbished soon. It will be a flying garden and kitchen, and a vehicle of social action. In a couple of weeks, we will be moving out of the campsite. Packing things up is part of our essence. We are nomadic. We move and adapt.*

*There are grey buildings around me as I write, around the loud Euston Road. Bayo Akomolafe[xiv], thinker and poet, says that it is important to hug monsters. I tried to avoid cold corporate buildings for a long time, instead staying in the Skip Garden to smell the herbs and listen to the bees. Now I am ready to hug this commercial campus, and try to help nature and young people to flourish, and interact with this 'monster' too.*

*We have a kitchen, an office space, a toilet, and a dishwasher, which is much more than what was available to us when I first started as an intern for Global Generation. Hopefully, this will be the first step of another long journey. Young people will be my companions, alongside all the great human beings I have met so far at Regent's Place.*

Mariam Hassan and children from the Surma Centre in the Milk Float which is inside our first urban campsite – photographer Jan Kattein

Children from the Surma Centre design the transformation of the Milk Float – photographer Jan Kattein

A transformed milk float 2019 – photographer Jan Kattein

Rod and Jane cooking off the Milk Float at the Caversham Medical Centre 2019 – photographer Jan Kattein

In order to help grow relationships in the local area, we paid Mariam Hassam, a local youth worker to run the Generator sessions with Silvia. Mariam worked for the nearby Surma Centre; (set up by the Bengali Workers association). She had been involved on the Regent's Park housing estate, situated on the northern edge of Regent's Place, for over 12 years and knew many of the children who lived in the area. She described how the experience of working with Global Generation made her look at nature and gardening in a totally different way;

> I have found a new passion and interest for nature. I love that I can see furniture being made right in front of my eyes by young people who are initially afraid to do so but feel so alive once they try it out. Watching the young people grow and leave their comfort zones, challenging themselves and seeing them enjoy learning and making new things is special to witness. The best thing that has come out of the sessions so far is what we don't really see from the outside but what the young people have taken in and what they have kept and used in their day-to-day life, and that is positivity and the willingness to try new things. Working with nature is not much of a draw card for many of the young people Global Generation is involved with; we have got used to the fact that initial recruitment onto new projects can be hard. However once started, they usually stay on board.
>
> Mariam Hassam – 10th July, 2019

## Bounded freedom

Soon we began hearing about ideas for greening the campus from the estates team; a garden made of recycled pallets in the plaza, soil supported by coffee grounds and perhaps a market. Whilst we could take no credit for the ideas, nor were any of them generated by the wider public realm master planning process, I was happy things were moving in the direction of initiative and sustainability. I was behind what I heard and keen to collaborate and support where we could. However, these grassroots plans were not received so well at a senior level. I happened to be sitting with Juliette when the email came in describing the estates team's ideas. To my surprise she wasn't as enthusiastic as I was. "What about the branding guidelines?" she said. With good intent and an in-depth process of consultation, British Land had come up with a new brand. The brand spoke to everything my colleagues and I felt strongly about; the environment, responsibility towards the local community, heart and soul. I naively assumed this might bring a greater degree of freedom and flexibility. However, I had completely underestimated the power and potential limitations of a brand if held too tightly. Where there had been an outmoded, loosely held outline, now there were very clear guidelines, which I am sure had not come cheap.

As we had hoped, pretty soon requests to Global Generation to design, deliver and maintain planters in the outdoor space came from the occupiers of Regent's Place, but actually doing this was not straight forward. Our first commission was to be a group of simple planters made out of scaffold boards, outside the Old Diorama Arts Centre, paid for by the centre. The plan was to do the planting with some of the children we had been working with. Out of courtesy more than anything else, I shared the ideas for our first commission with the Regent's Place estates team and marketing teams. I also met with the same response: "How will they fit with the brand guidelines? Have you thought of incorporating our new R (for Regent's Place) into the design?" This response triggered a lengthy process of approvals and in my mind unnecessary layers of work. Firstly, producing a drawing that might be accepted by multiple departments and various different agencies involved in the delivery of the brand would require a proposal in a form they would recognise. This meant asking our architect to produce a suitable document and so began a complicated back and forth to achieve something which literally could have been done on the back of an envelope. After all, we were only talking about three planters made of scaffold boards. Luckily, the Director from the Old Diorama was understanding and added more to our fee. He had spent five years unsuccessfully trying to get consent for planters outside the Arts Centre, so he knew what we were up against. The weeks went on and our plans bounced from one desk to another, with no forward traction. Fortunately, I had a catch-up meeting scheduled with Juliette. As with most of our meetings, we didn't set a formal agenda in advance, but I had some key points I wanted us to discuss and so I wrote them in a little notebook. In the margin, in my fairly illegible handwriting, I wrote five words. As our time drew to an end, Juliette looked down at my list, and her eyes settled on those five pivotal words, "The tyranny of a brand." As I remember it, there was little conversation needed, she just looked at me and said, "I got it." That afternoon I received a phone call from the Brand & Customer Marketing Manager for Regent's Place, who seemed genuinely happy to inform me that she just had a call with Juliette and they had taken a U-turn. They felt they should hold the brand lightly, and in that spirit, diversity through a variety of planters on the campus could work really well.

## Reflections

Action research encourages a participatory approach, whether that be in the creation and delivery of a project or how the story of a project is told. Writing this chapter occurred over the course of a year through the gathering of stories from the Global Generation team who worked at Regent's Place. These stories came in the form of blogs, journals, conversations around the lunch table and emails that came our way. Writing about tensions and hidden dynamics was sustaining; not just for me but for a number of my colleagues. The process of finding words to capture our experience, lifted us out of the challenges that arose in pioneering new work in an environment which was not set up for our way of working. Silvia summed this up in a poignant blog post; which was well received by British Land:

> If you ask me how I feel, I don't exactly know. Perhaps I struggle to embrace the predominant culture that surrounds our campsite at

Regent's Place; a culture that seeks order and control. Is it because our fingernails and our floors might get dirty, that sometimes it is hard to hear a hello and I feel unwelcome?

I love my job, I love what I do and I believe in the fusion of nature and young people as a key for social impact, but there is uncertainty in Global Generation's new place to work. Is it the weight of bureaucratic rules imposed from elsewhere that makes me feel a bit trapped and foggy, struggling to express the best of myself? Is it the need for certainty and perfection that keeps pushing away the wild? But despite these moments of vulnerability, I am enthused by the pulsing life and beauty created by the young people who have come to join us.

Silvia Pedretti, 24th June, 2019

It is all too easy to slate the mechanistic approach of modernity. In finding our way forward from the environmental mess we have created, what are we to do? I appreciate the proposition put forth by Feinstein et al, which speaks to the aspiration for Global Generation's work; "The ideal alchemy retains the strengths of both the prevailing myth and the counter-myth while transcending their limitations. Sustainability does not abandon but rather reconciles itself with progress."[xv]

In compiling the writing for this chapter, the aim was to give a flavour of lived experience, the rough and the smooth of it all, whilst honouring the many people who have helped us. I see Global Generation's role in a corporate environment like Regent's Place as that of being a critical friend and we had a lot to learn. We sat on the outside of the inside and were happy collaborators. The biggest learning was how much work and patience was needed to launch a project within a corporate world where health and safety and reputation around market value was a very real concern. Stepping out of our Skip Garden compound and agreeing to work on the inside of Regent's Place exposed us to more rigid structures than we had encountered before. In this process, time was our most valuable friend. Over time, rules and role identities dissipated and more intuitive and trusting ways of knowing became available. Freya Mathews describes a process of what she calls "re-inhabiting reality." I take this to mean calling forth and making space for a more original reality; the wild that lives beneath the concrete in places and in people, whilst finding a way to understand and work with all that we have become. Mathews' words quoted earlier acted as a helpful touchstone: "To re-inhabit the places in which we live is not to raze the smokestacks and freeways that we might find there but to fit them back into the larger unfolding of land and cosmos.[xvi]

For our contribution to be grounded and locally relevant it was important that we slowly felt our way in, listening for what emerged in our conversations with each other and with the people who lived and worked in and around Regent's Place. Eventually, barriers disappeared and goodwill and common sense found a way through. Six months after the emails requiring risk assessments for anything that would move, our huge basement campsite was a very different place. On one memorable occasion when I was conducting a job interview at one end of the campsite, the place was humming with the sound of young people at the other end, along with a quiet but continuous beeping. One corner of the campsite had been converted into a wood workshop and although the dust had gotten into the alarm system, the estates team were aware of it and were relaxed. Children and young people were busy preparing for a Christmas market. In the courtyard, a young boy wearing a face mask and ear muffs held a massive sander. He was turning cheesed logs from Hampstead Heath into chopping boards, while others were busy with saws making candle holders and in the kitchen jams and chutneys were brewing. None of the activities required special permissions.

I have often said that for organisations like Global Generation working in places like Regent's Place it is important not to be extensively funded by a commercial marketing budget. The marketeer's need to make things happen quickly with compelling stories is different to the slower, more organic rhythms our work has thrived on. In terms of an approach to marketing, Global Generation has historically found excuses to begin with live action on the ground and then we have shared the stories of what happened. In most commercial marketing campaigns, the narrative of what might be is often told first as a way of drawing people into an area. I realise that I had followed the same tactic in writing and sharing the story of the Wise Witch. I learned that storytelling used like this is a powerful way of evoking an alternative vision of what might be. As Alex Evans, who wrote The Myth Gap,[xvii] suggests, if you want to change the world, more than facts, what you need most is a really great story. At the same time, this is an approach that should be held lightly and critically. Evans also points out; to our peril Donald Trump and Nigel Farage triumphed through their storytelling abilities rather than their evidence base.

We did our best to provide inspiring and factual content in the conventional form of reports, proposals, and PowerPointed plans. Whilst it was a story I seldom shared, in my mind's eye, I kept coming back to the mythic styled story about the wise witch overcoming the ogres, with one foot on the shiny granite and one foot in the belly of the earth. This story would re-surface at unexpected moments, acting as a kind of business plan, a hopeful reference point from which the material realities of our work could unfold.

True to any good collaboration, working at Regent's Place has been a delicate balance of striving to have our own way and learning how and when to follow the rules. Maintaining honest dialogue with the community engagement, marketing, estates, sustainability, and asset teams has helped grow respect and integrity in our practice. It has helped me see and appreciate people and the challenges of their jobs. An example of the importance of dialogue is the way this chapter was written. I am an insider researcher, and unlike many studies, I have chosen the somewhat difficult path of naming many of the people and the organisations that I write about, hence it was imperative that I share an early draft with Juliette. She gently pointed out that whilst at times we may have experienced Regent's

Place as unfriendly and full of rules, we had a lot of support from within and it was important to acknowledge this. She was right, it has been the work of a thousand hands, many of them unseen. `Our projects would never have flourished if we had to push our way in, it has been a mutual endeavour.

## Notes

i Shaw, M. (2014). *Snowy Tower: Parzival and the Wet, Black Branch of Language.* Ashland, OR: White Cloud Press. p.153.

ii In terms of what is offered as reality in today's mechanistic world some may consider the enchantment insubstantial or representative of primitive magical thinking. As you will have experienced in previous chapters, I have come to appreciate enchantment as neither of those, nor as an idealised state, but as that; indefinable, unbroken essence that gives life in whatever form it takes, its deepest meaning. I feel that enchantment is a condition/world that exists despite our dominant world. Enchantment runs beneath the fragmentation of the mechanical mind and a colonialist past. Naming and specifically defining this older, deeper, richer, more connected sense of self seems impossible. History would suggest that naming runs the risk of owning and fencing as opposed to discovering freshly each time.

Curry, P. (2012). Enchantment and Modernity. *PAN: Philosophy, Activism, Nature 9,* 2012, pp.76-89.

Mathews, F. (2003). *For Love of Matter: A Contemporary Panpsychism.* Albany, NY: State University of New York Press.

iii Johnston Phelps, E. (1993) *Tatterhood and Other Tales.* New York: The Feminist Press at CUNY (first

iv Aubrey, N. "'A Dwelling Place for Dragons': Wild Places in Mythology and Folklore." *The Psychology of Religion and Place.* Palgrave Macmillan, Cham, 2019. 145-166.

v Price, T. (2014). *Why Leaders Need Not be Moral Saints.* In: J. Ciulla, ed. *Ethics: The Heart of Leadership* Westport, CT: Praeger Quorum, p.133).

vi Berry, T. (1982). *Teilhard in the Ecological Age,* American Teilhard Association, *Teilhard Study # 7,* Fall 1982.

vii King, U. (2000). `Rediscovering Fire - Religion, Science and Mysticism in Teilhard de Chardin. *Earthlight magazine,* issue 39. p.44).

viii Gregory, J. (2008). PhD, *Ethics and Professional Practice* (unpublished).

ix Lee, M. (2003). On Codes of Ethics, the Individual and Performance. *Performance Improvement Quarterly,* 16 (2) pp. 72-89.

x Latour, B. (1991). *We Have Never Been Modern.* C. Porter, (Trans 1993). Cambridge: Harvard University Press and Curry, P. (2012). Enchantment and Modernity. *PAN: Philosophy, Activism, Nature 9,* 2012, pp. 76-89.

xi Curry, P. (2012). Ibid p.76).

xii Cartesian refers to Rene Descartes who described nature as a great unfeeling machine, strictly arranged by mathematical law and controlled by an external creator. This created the psycho-physical split between mind and matter known as the Cartesian split. Descartes crystallised the notion that man and nature were two separate things and in this light a meaningful cosmos faded. We lost our ecology and our cosmology and in that sense we also lost our story. With no sense of intelligent life left, other than in the mind of the privileged human, our job was to command and control. See: Griffin, D.R. (1993). Introduction: Constructive Post Modern Philosophy. In: D.R. Griffin, J.B. Cobb, Jr, M.P. Ford, P.A.Y. Gunter, and P. Oches, eds. *Founders of Constructive Postmodern Philosophy.* Albany, NY: State University of New York Press. pp. 1-42.

Tarnas, R. (1991). *The Passion of the Western Mind: Understanding the Ideas that have shaped our world view.* London: Pimlico.

Toulmin, S. (1985). *The Return to Cosmology: Postmodern Science and the Theology of Nature.* Berkeley, CA: University of California Press.

Worster, D. (1977). *Nature's Economy: A History of Ecological Ideas.* Second Edition (1994). Cambridge: Cambridge University Press.

xiii Krishnamurti, J. (2010). *Freedom from the Known.* London: Rider Books. p.127.

xiv Akomolafe, B. (2017). These Wilds Beyond Our Fences. Berkely: North Atlantic Books

xv Feinstein, D., Krippner, S., and Mortifee, S. (1998). Waking to the Rhythm of a new myth: Mythic Perspectives for a World in Distress, *World Futures.* 52, 187-238. [Online]. Energy Psychology. Available at http://www.innersource.net/ep/images/stories/downloads/World-Futures-Article.pdf [Accessed: 23 April 2020]. pp. 187-238.

xvi Mathews, F. (2003). *For Love of Matter: A Contemporary Panpsychism.* Albany, NY: State University of New York Press. p.200).

xvii Evans, A. (2017). *The Myth Gap: What Happens When Evidence and Arguments Aren't Enough* Cornwall: Eden Project Books

# 10

# FROM THE HEART OF AN OAK FOREST

> To be truly free one must take on the basic conditions as they are – painful,
> impermanent, open, imperfect – and then be grateful for impermanence
> and the freedom it grants us. For in a fixed universe there would be no
> freedom.
>
> Gary Snyder, 1990.[i]

Time was running out and Global Generation would need to move from the
Skip Garden in 12 months' time. This chapter describes some of the leadership
dilemmas I encountered in this next step for Global Generation. I describe elements of what it took to make a community garden happen in Central London
in terms of the nuts and bolts of finding a site and gaining permissions from large
institutions, consent from the local community, and creating an infrastructure
to operate from. These were the front-facing aspects of bringing our new home
to life. However, there was a less visible underbelly that arose in response to
the uncertainties of this next move. Because this new project was larger than
anything Global Generation had done before, it gave rise to anxiety in some
quarters about our unconventional, wayfinding style of doing things. We lived
to tell the story without any major falling out, and in my view, became stronger
and clearer about our organisational culture because of it. To make a new community garden a reality, it took a web of relationships, complex agreements and
if it were not for email, a small forest of paper. Unlike the development of the
previous Skip Garden's, it wasn't me, but Nicole van den Eijnde (who was now
joint director of GG), who led the back-end negotiations needed to make this
new garden a reality.

The stories I tell are largely from my perspective and, as with previous accounts,
they are partial. There are many more stories that could have been told; far too
many for the constraints of this one chapter which means that unfortunately

I have not been able to do justice to the creative contribution of the many people who made this next step for Global Generation a success. This includes the fact that in many ways we stood on the sometimes invisible shoulders of individuals and organisations in the local area who had paved the way for us. I have chosen to focus on the navigation needed to achieve a largely participatory process, in a time-limited context with multiple and diverse stakeholders including our trustees, local residents, professional associates, and partnering institutions. To bring what are sometimes separate worlds together, we needed to create the conditions for different types of knowledge and experience to be valued and made visible. It was a process that involved working with the inevitable tensions of combining different ambitions, skill sets, and ways of working; a process in which we didn't always agree. I have also tried to surface the often buried constraints inherent in the legal and cultural structures of a charitable organisation, where ultimate responsibility sits with those who are not immediately involved in the hands-on work on the ground. The wider context was that the world had changed since the establishment of the first Skip Garden in 2009. The coronavirus pandemic was waiting in the wings, global warming was a reality, and the question of how we could make a positive contribution in a climate-changed world pressed upon us.

## Where to next

New social housing units were scheduled to be built on the land where the Skip Garden was based and despite months of effort to find a suitable, rent-free, vacant piece of land, we had nowhere confirmed to go and the clock was ticking. This was not simply a question of moving six skips and a gazebo as in the early days, but a team of 15 staff and an extensive education and events programme, along with a back-end for commercial and community garden commissions and a café. The King's Cross estate was closing in on us, multiple tall buildings rising up on every side of the Skip Garden, and all available space was at a premium. Argent, the developers, were keen to lease Global Generation a permanent site for a peppercorn rental. This was an area of land deemed by the local authority as an ecology site and a once in a lifetime offer; however, this was some years off. Instead, as an interim measure, we were offered a shady strip of land next to one of Google's artificial intelligence companies, which inspired none of us.

Our success had always depended on relationships with the schools, young people, and community groups surrounding the Skip Garden, as well as relationships with the construction companies working in the area. These connections had been built over many years, meaning that moving our operations to another part of London was not an option. My walks to work each morning were taken up with imagining how we might occupy various plots of vacant space encountered on my way: the siding of railway tracks, a part of the canal, an electricity substation, an old petrol station, a disused tube station. Finally, we identified a site: just over two acres of land that looked like it might be available for several years. The land belonged to the British Library and was a perfect location; this was yet another "meanwhile" site that would not be built on for three years or more.

Turning a possibility into a reality that we felt comfortable inhabiting was a delicate balancing act which raised many questions. How much of a plan did we need to set out in advance? How much could be left to emerge between the people who came to work in the garden? On the other hand, to raise funds and to even be considered a viable occupant by the powers that be, we needed to show in a convincing and inspiring way what might be possible. We drew on the expertise of our architect friend and associate, Jan Kattein; having led the build of the last Skip Garden and projects at Regent's Place, Jan knew our practice well. We also knew that, opposed to coming up with a rigid master plan, Jan is an enabler committed to a "commonplan" process. Jan describes how a commonplan allows for the kind of unexpected and organic processes that begin to happen when people are able to work together and over time continue the physical transformation themselves.

> The very term – masterplan – suggests a masterly and exclusive endeavour. A commonplan in contrast is an open and inclusive framework that draws on input from many stakeholders. As a result, there is not one single author, but authorship is diverse. A commonplan gives form and shape to a vision owned by many. In the case of the Story Garden, we knew that some basic infra-structure would be needed in order to make the site and project accessible to others. The commonplan proposed a half-developed garden with the essential infra-structure in place, but it also provided for future change, adaptation, and customisation by others.
>
> Jan Kattein, May, 2020.

Jan, Nicole, our chair of trustees, and I presented the plan to a team from the British Library and Stanhope, the company who would eventually be developing the land for its permanent use in a few years' time, after the "meanwhile" use period had finished. We hoped for both approval to lease the land for no rent and at least some of the funding that would be required to transform it into a community garden. Whilst nothing was agreed around the rather formidable boardroom table, we left feeling optimistic.

All parties were keen to know what the people living around the site in Somers Town wanted from the garden. However, consultation was tricky in an area that has arguably seen more changes and is populated with more universities than any part of London. Consequently, local people had been "over consulted" with very little to show for it. Like any city centre, land was at a premium and there were different ideas about what should be done with the site; a garden, a dog-walking park, a car park, a container village of business start-ups ... The list went on. Global Generation had worked in and around the proximity known as Somers Town for more than 15 years and some of us lived not far away. That being said none of us actually lived in Somers Town. We were aware that despite our long-term local involvement, Global Generation could still be perceived as a privileged group of outsiders without any real local commitment. Hence our interventions needed to be light, inviting, and meaningful. As much as possible, we needed to grow the garden with the people who lived in the streets surrounding the garden. The photograph below shows how the Story Garden started as a barren site that could be viewed from the outside on three sides. Unlike the Paper Garden, which is tucked away and not visible to the local community, this vacant site was highly visible; flanked to the south by the British Library, to the west by Levita House, a large housing estate, to the north by the Francis Crick Institute, and to the west by St Pancras Station. This garden would also be open to the public from the get-go.

The beginnings of our new garden with the Francis Crick Institute in the background – photograph by Jane Riddiford

We decided it would not be helpful to call this new site the Skip Garden. As mentioned previously, it was an accolade that, at least in London, the Skip Garden became a symbol of community gardening, focusing on the Skips alone missed the point of collaborative involvement with local people. It needed to reflect the immediate locality and we wanted to know if and how local people would like to use the garden and what they thought the garden should be called. Many names were discussed and voted on, including a Woven Garden, a Seedling Garden, and a Library Garden. In the end, the Story Garden proved to be the most popular name for our new home. As well as being close to a famous library and a home for our growing practice of storytelling, the first months of bringing an empty brownfield site to life were a story-making process. Through hands on involvement, together we began to grow a living story. The two photographs below were taken on a hot July Saturday when we opened the garden with a day full of activities, including seed sowing, carpentry, the big dig, illustrations from Somers Town artists, and a pay-as-you-feel lunch. We were over the moon to feel the garden becoming an exciting community space.

Story garden hoarding summer 2019 – photographer Jan Kattein

The story garden opening July 2019 – photographer Adam Razvi

The garden was more than twice the size of the Skip Garden, nearly two acres and we were able to introduce new elements into what was offered. A partnership with the University of the Arts (UAL) involved them in developing a Maker Space complete with sowing machines, wood-cutting equipment and a community kiln. This was a space for students and staff of the university to work more directly with the local community. For the first time ever, we were able to provide a local grow bed scheme, which is pictured below. There was not enough land for everyone to have their own plot, but an arrangement was settled on whereby local people had to put themselves into groups in order to have a grow bed allocated to them. Each week on a Thursday evening, we continued our practice of Twilight gardening. After work local residents and employees helped us make wooden planters and fill them with soil. This was not without set-backs, as we discovered the six tonnes of soil we bought in were filled with herbicide and the plants noticeably struggled. Without missing a heart-beat volunteers pitched in, and over two weeks the soil in each growing bed was replaced.

Story garden community grow bed holder during the COVID-19 pandemic with St Pancras Station and the British Library in the Background – photographer Kiloran Ben O'Leary

Each week, family Saturdays were hosted and during the week, volunteer sessions, school workshops and stay and play sessions for nursery children were a regular feature. This is how one of our community gardeners described her experience:

> We sowed broad bean seeds, transplanted winter salad, created clay tree faces, told stories, ate a great deal of snacks, splashed in puddles, got inevitably wet while watering the garden, harvested tomatoes from the poly tunnel, learnt how to use a knife in the kitchen, laughed a lot, and had many rides in the trolley we usually reserve for moving heavy items around the site! Smiles were infectious, splashing in puddles were a must and parents and children opened themselves to the fun of being outside in the elements. As a facilitator, I witnessed children grow in confidence in their surroundings and themselves.
>
> These sessions bring a certain magic with them. It is a time for true play and the freedom that comes with it. Want to play with clay? Rather run around in puddles and sing? Great! It is also a time for exploration and taking time to explore the smaller aspects of the garden. What does this leaf feel like? Does it feel soft or rough? Can we see a shiny leaf? Can we see a purple flower, does it smell?
>
> Charlotte Gordon, GG Community Gardener, December 2019

## Finding our own feet

It was several months after planning consent had been granted and the contractor had completed most of the basic infrastructure stages of the build that we really got to grips with the detail of the designs we had presented to the British Library. Understanding something on paper is very different than seeing it take shape on the ground. The challenge is that conventional design and construction processes in the UK are strictly sequential, construction-related health and safety legislation is rigid and by the time that works start on site, we realised that it becomes more difficult and costly to make changes. Despite the inherent flexibility of the commonplan that Jan had worked up and the best intentions all around, the reality of passing the baton of responsibility from one hand to another was a little rocky.

Unlike the previous Skip Garden, this was a large site and easily partitioned which meant our contractor (found by Jan, approved by the British Library and appointed by Global Generation) was comfortable having community planting and volunteer involvement run alongside the construction work. On many sites, this would not be an option, due to health and safety and liability issues. As funding bids from Local Government and various trusts were successful, we were able to appoint Kiloran Benn O'Leary as our Gardens Manager, Charlotte Gordon as a Community Gardener, and Martina Mina as Community Build Manager. This enabled us to have a five-day presence on the ground. Each week, hundreds of volunteers made planters, moved barrows full of soil and woodchip, and planted herbs and vegetables along with nurturing the nettles, the mugwort, and all the many other wild volunteers that had arrived well before us. Excitement was running high as the garden took shape around us. Even though containers had been used, Jan and his team had successfully avoided the all too common box park effect. The designs that had been submitted for planning included a wide, angular, black-brick and rather costly path, that would run around the site linking and grounding the various structures; it was in the contractor's brief to lay the path. The bricks for the path were sourced, lined up and work was to begin the next day. However, the GG Story Garden team, who now considered the garden home, suddenly decided they didn't want it that way and at the last minute put a stop to the works. They wanted to use natural materials and curves and most importantly we all needed time to live our way into the design and flow of the garden. Though it might sound like a small thing, it was a significant moment in which members of the GG team found their own feet and took on real authorship for the site and the energy in the garden took a leap forward. The decision to abandon the brick path was made at short notice, and despite it being inconvenient for the contractor and on a design level disappointing for Jan, they both stood back. We wanted to encourage people to go off the path and as the garden grew, the plants became the social space whereas Jan had felt the path would be the central and most important social space.

> The intention of the path was to create a (pedestrian)"street" as part of the basic infra-structure that we were providing for the site. This street – so I thought – would anchor the buildings and spaces on either side and leave you with a series of pockets that you could inhabit, adapt, and customise as you continued building the site. Streets (going right back to roman times and beyond) for me are also some of the most important social spaces that anyone had ever conceived. Along them grew markets, out of markets grew villages and towns. Removing the street suddenly when it had been there all along was a disappointment for me seeing that I understood it as one of the most important "seeds" that we had sown.
>
> Jan Kattein, May 2020.

In terms of functionality, especially access for buggies and wheelchairs, abandoning the brick path at this early stage was a difficulty. In one way or another, I have been writing about "the enduring hostility between the institutional order and the forests that lie at their boundaries."[ii] Whilst not articulated at the time, it seems that rather than the civic heart of the Romans, the Story Garden team were pulled by the wild heart of the forest. For a smoother ride, we might have discussed this from the outset, but primordial impulses are seldom revealed that way. We did eventually achieve a path, which began with re-used scaffold boards. As time went on, the whole shape of the garden evolved into a series of concentric circles as opposed to straight lines. Changing the path, re-positioning the orientation of a two-story portacabin classroom, and putting additional windows in our upstairs container office (in order to gain sight of visitors coming in the entrance gate) were some of the things that were able to evolve from what was set out in the planning permission document.

## Behind the scenes

In many ways, the Story Garden was a useful test bed for longer term opportunities that would be coming our way. We were now an organisation with 15 staff, public realm commissions, and a future offer of taking on a permanent rent-free site with a 99-year lease. Understandably, our trustees worried whether our staff team had the expertise to identify and overcome things that could go horribly wrong, and if we would have sufficient funds to rectify them. Coupled with this, the leadership style that Nicole and I were committed to, in terms of wayfinding and sharing responsibility within the whole team, was hard to understand for some members of the board who were used to more conventional ways of running an organisation. As one of them let slip, "It is all very well these fluffy ways of working, but now we have to operate like a real business."

The first part of the Story Garden build was well underway, but there was still a lot to do for it to become a home that could accommodate staff, volunteers and our different education programmes. We had an ambitious plan which would stretch our ingenuity to the limit. Neither Nicole nor I saw any reason to abandon our collaborative ways of working; now was a time when we needed everyone's input. We were encouraged by the growing capacity we were witnessing throughout the existing staff team, from the most junior person upwards, along with a wider body of committed associates and volunteers. Most of them were women, they didn't lead in a particularly visible way, but rather built relationships and just got on with whatever needed to be done. We felt our hidden resources, in other words, the strengths of the quieter voices both within the staff and the trustee board were not being taken into account by some of the more vocal trustees in their repeated requests for us to employ someone more experienced; someone at an "executive level" as they put it. Along with the shortfalls on our budget, Nicole and I were concerned about undermining the team by helicoptering in expensive help from the outside, preferring to grow the team slowly from the inside. Some trustees felt we needed someone who could play hardball in commercial contexts; I felt sure our commercial success was based on our relational approach. Some felt it wasn't necessary for all of the staff to align with our values and ways of working, whilst others felt this was critical.

Conversations were underway about how the rest of the move and the second part of the Story Garden build would happen but nothing was confirmed. I felt a quiet confidence that it would all be done in the same way it had always happened, through the in-kind support of the different King's Cross construction companies. It was now questioned by at least one trustee whether or not as a larger organisation we should be relying on pro-bono help at all. In one memorable meeting, an industry project manager was brought in. He came up with a £100,000 estimate for work, much of which we had done twice before at no cost. This was followed by more insistence that we should employ executive support. I tried to explain that we needed time to feel our way into what was needed, in our own shoes, but my words went nowhere. I was incensed and concerned that our previous experience and hard-earned ways of working seemed to count for nothing. It felt like the heart of the organisation was being ripped out. Untypically for me, I exploded and in a very loud voice said, "No, No, No!!!" Around the table there was a stunned silence. Everyone was shocked, including me. In that moment, I wanted to run but I didn't, I stayed put, calmed down and apologised for raising my voice. At the same time, it felt clean and right to have put my foot down.

Not long after, Nicole and I received a very reasonable email from the trustee who had been in my line of fire, setting out how "we needed to have a stronger executive management structure within the charity to manage the day-to-day administration that is a pre-requisite of working in the modern environment." He went on to say that he didn't feel this view was universally supported and rather than get into conflict he would like to step down with immediate effect. After reading the email, I thought to myself, perhaps the sticking point is that we are operating with post-modern sensibilities, where there are not such obvious demarcations of power and hierarchy. This can allow more room to grow skills and strengths across the team and invite unexpected possibilities, but this can also be confusing, frightening, and run the risk of no one taking responsibility. This is even more intense when reputations are at stake. I empathised with why the trustees might be concerned. Nicole and I had a lot on our plates; they worried that we would run out of steam and leave and they would then be responsible for an organisation whose ways of working some of them struggled to understand. The previous year I had had a cancer scare, and I wondered if this was playing a part in the growing level of anxiety that was permeating the board. Thanks to the gentle and thoughtful guidance of two of the trustees, Nicole and I began to work out where we needed help, but it would take many more months before we were ready to advertise for a new role. Whilst the backroom dramas were bubbling away, things were quietly moving forward in the Story Garden.

## A tree sanctuary

Global Generation's sense of community has been shaped by ecologist Aldo Leopold. In 1949 in his legendary book, "A Sand County Almanac" he described how community is not just about relationships between different people but between all of us, soils, waters, plants, and animals. Not only was the Story Garden to become a refuge for the local community but very soon it became a sanctuary for trees. An email arrived from our new friend and collaborator, Polly

Gifford of *Theatre Complicité*. The email heralded the arrival of a portable wood: a band of 52 trees with a message, coming to the Story Garden.

> I would like to introduce you to Heather Ackroyd and Dan Harvey, fantastic artists and part of the original Culture Declares Emergency group. They have a wonderful project called Beuys Acorns currently installed in the Bloomberg Arcade in London. They might need to look for somewhere to keep the trees from September, I had spoken about Global Generation previously and said I would make an introduction in case there was any possibility that the Story Garden could be an option. The trees are very evocative and themselves full of stories.
>
> Polly Gifford, August 2019

In 2007, Heather and Dan (known as Ackroyd and Harvey) gathered and germinated hundreds of acorns from renowned artist Joseph Beuys's seminal artwork, 7000 Oaks in Germany, and in doing so began a new open-ended research project which they refer to as Beuys' Acorns. In the 1980s, when the original 7000 Oaks were planted by Joseph Beuys, they held a message of social and environmental change; a message the Oaks carried with them into the Story Garden. The reason why they needed a home in London was in itself a call to action, speaking to the wider problem of importing mature trees, a common practice in new regeneration schemes and a potential factor in the migration of plant disease. Due to the spread of processionary oak moth in London boroughs, there were now limitations on where oaks of any description could travel. Whilst the Beuys' Acorns had a clean bill of health, their presence was a poignant reminder of the growing number of pathogens that are endangering many plants, like the beloved ash in the nearby St. Pancras Gardens, otherwise known as the Thomas Hardy Tree.

In the spirit of the Story Garden, the Oaks were arranged in a circle of three rings of trees, with a fire pit and storytelling space in the middle. A circle that would spread across the space where the brick path would have been. We noticed that people were instantly drawn to be with the trees. In the first week, the Oak Circle was home to a talk for a group of volunteers by Sue Amos, GG's new head of gardens, about the importance of bees and other pollinators in the city. The volunteers created a wildlife corner in the garden. Two days later, our family day participants sat amongst the Oaks. Along with storytelling, fire lighting, and the cooking of thyme tea on the fire, the mums told us stories of family picnics in Bangladesh. They asked if they could cook for us and if we could run an overnight camp in the Story Garden for them and their children. Within the call for positive social and environmental change, the Oak Forest carried a message of safety, intimacy, and healing. As one mum said; "This a place we can be and mix with each other … I feel good in here."

The Oaks (pictured in the photograph on next page) were soon followed by a gift from the Green Legacy Hiroshima project which was established to safeguard the *hibakujumoku* (trees that survived the atomic bomb). A central part of the initiative was sending survivor seeds to over 30 countries and 1000s of

From the heart of the oak forest – photographer Silvia Pedretti

projects. Thanks to local resident, Apolonia Lobo, we were fortunate enough to receive seeds from the survivor trees so that that we could plant trees as a symbol of the kind of future we wanted our new garden to support.

## More behind the scenes

Our chair of Trustees, John Nugent, had said he would be stepping down and this introduced a new topic into our board discussions. Tony Buckland, our former chair, had recently re-joined the board as a trustee and he agreed to help recruit John's successor. Over a lunchtime meeting with Nicole and I, Tony had suggested appointing someone who understood our philosophy. We all felt that this person might be hard to find if we advertised the role and that we should draw on people within the board and our shared contacts. Tony raised the point that the traditional charity structure, where legal responsibility for big decisions sits with the trustees not the executive, may not be an ideal fit for the more collaborative culture of Global Generation. I agreed and added that it can be particularly difficult for any organisation where the founders become paid executive officers and therefore in most cases, they are not allowed to be trustees. This is one of the reasons why some organisations decide not to become charities. However, I felt that so long as we were legally compliant on what we must do in terms of charity and company law, we needn't be bound by recommendations, which in my view seemed outmoded. In other words, Nicole and I could still work closely with the board in all decision-making, unless of course there was an issue of serious concern, then full responsibility would revert to the trustees.

A few weeks later Tony and I arranged to have a phone call. During the call he said that he had had been thinking that perhaps we should appoint someone new and completely unknown to Global Generation to be chair of trustees. Much later, Tony explained his thinking to me.

> I was trying at that time to deal with the fact that the research I had done into the appointment of a Chair and trustees of charities, had shown me that maybe we should approach things slightly differently. The appointment of trustees and a Chair is the responsibility of the trustees and indeed the trustees make that appointment. In some examples, I researched the Executives were largely excluded from that process. That was not what I was suggesting but I began to wonder. Hence, I suggested that on reflection a chair from outside might be worth considering. The Executive (i.e. Nicole and you) had always been intimately involved in the selection of trustees and past Chairs at GG and I was not suggesting that you should not be involved. Rather than exclude you both, I was trying to ensure that the trustees were aware that the decision was theirs to make.
>
> Tony Buckland, April 2020.

In hindsight, I can see that at the time I drew more extreme conclusions than Tony was implying. The fact was, I was still raw about the difficult trustee dynamics of the preceding months. I remember listening to Tony and feeling quite deflated; just at the point it seemed like we were getting somewhere that we could all agree on, it was disheartening that we might consider going back to square one and deliberately seek out someone unknown to Global Generation. The idea that in some circumstances Trustees might make a decision like that without the executive sounded retrogressive and even draconian. I thought about our staff appointments where we had made extra effort to give the people on the ground an involvement in the interviews and a final say about who they thought they could work with. I was used to feeling my way into what might work with Tony and the other trustees rather than being told what to do; choosing a chair had always been something we had done together. Anger and resistance to suggestions that I perceived to be borne of a world where dominating control and patriarchal power is the norm began to brew inside me.

## My Journal

*Last night I tossed and turned, sleeping and waking, turning over in my mind the phone call I had with Tony. The conversation took me by surprise, it seemed like a U-turn on what we agreed over lunch a few weeks ago. I can't understand the logic or the business sense of it all, in my view there is nothing that requires a change in the way we work; nothing is broken, despite a dismal financial forecast some months ago, the figures are now good, despite some anxiety about our capacity to make the transition to the new Story Garden, all is going well, in fact beyond expectations.*

24th Sept, 2019

As I lay in bed my thoughts were wild and untenable. I began to wonder how you would know whether or not you were living in the middle of a coup. What I couldn't understand was that there was no visible problem or reason to change. To me the problem seemed to be fear amongst some members of the board about what "might" go wrong rather than what was going wrong. In the past as our chair Tony was up for embracing new leadership ideas; he had enthusiastically taken part in Chellie Spiller's Wayfinding Leadership workshop and the workshop with Mary Evelyn Tucker on the Universe Story. However, he had recently said on more than one occasion that his reference point was the responsibilities he had had in his former career with a bank. Amongst other things, he had been responsible for risk management. "We are not a bank," I thought. "Is this a microcosm of what is happening in our wider world?" In the face of uncertainty, some people lean into new and unexplored ways of doing things and others understandably default to the old and the familiar and the polarisation becomes crippling. And so, my thoughts went on as I tossed and turned. In desperation, I picked up a book by John O'Donohue called The Four Elements. I began reading about fire and boom that was it, a fresh perspective broke through in me.

> We live merely on the surface of the earth. Just a small distance below this surface, there is the raging fire of molten magma. At one level, it is a consoling thought that at its heart the earth is not cold. But at another level, it is quite a terrifying thought that there is fire burning at the heart of the earth. We forget this until a volcano erupts through the earth's surface. Then the hidden furnace at the centre rages forth. This recalls the ancient fire out of which the whole universe was generated.[iii]

I had the feeling that the fireworks I was experiencing in Global Generation were a good and a necessary thing, a creative friction out of which something new could be borne. This was it, "I could stand in the fire with my eyes open, I could appreciate the role of fire as a cleanser, as a healer, as a force of creativity." I got up and started writing and thinking about how we might start our next trustees meeting in a sacred and supportive way, opening the space for the appreciation and the uncertainty that exists between us.

Later in the morning, I met with Nicole to finish preparing our report for the next Trustees meeting. I appreciated her pragmatic, down-to-earth approach and could feel in myself that I had slowed down and moved into a more positive place. We both recognised that we needed to help the trustees and make it easy for them. We needed to explicitly respond to all the requests and commitments that were agreed at the last trustee meeting. As we talked, we realised we were not a million miles away from wanting what the trustees wanted. They wanted to help us but our way of doing things was different and unsettling for some. I rang Steve Marshall who was already on our board and who had expressed interest in becoming chair. I asked him to set out in a letter to the trustees what his motivation was and what he might have to offer. I noticed in speaking to him I was light and simple. I didn't want to burden him with my concerns. I did manage to explain where I thought difficulties and differences lay without it sounding like an obstacle.

Later in the day, Steve's email came through. He described his professional and charitable background and then went on to clearly set out his approach to leadership and management and his awareness of how difficult paradigmatic change can be.

> As Jane's previous doctoral supervisor, I share her values-led approach of participative, emancipatory, socially just leadership and management. I'm also deeply aware of the potential clash of paradigms that this can provoke and deal with this dynamic on a daily basis as a research supervisor and consultant in an organisation that is unashamedly commercial and transactional. My sense of GG in the future is that we should hold on to this hard-won value set and continually engage with the paradoxes and dilemmas this stance provokes in everyday, orthodox "business." I'm under no illusion about the anxiety incurred by this position and the "holding" that is required for all of us to work well together in this way – and realise that the weight of this responsibility lies predominantly with the Chair.
>
> In this respect, rather than any specific capabilities, I imagine that the most significant value that I could offer Board colleagues is broad experience of organisational development, dialogic and inquiry-based approaches to leadership and a passion for building community as a way to effect social change. In many ways, this position is the culmination of my "life's work" as I now try to bring more reflexive, considered ways of working into what I view as an increasingly complex and uncertain world.
>
> Importantly to me, I would only consider the role with express support from each of you; I really don't feel that leadership should be a win/lose battle. If I'm not the right person for the job, I'm fine with that. Obviously, there WILL be others "out there" and I will continue to enthusiastically support GG in the best way I can …
>
> Dr. Steve Marshall, September 2019.

A few days later, I was in the Skip Garden, about to depart after a three-day action research workshop, which was part of the MSc that Global Generation was delivering with Middlessex University. I said good bye to Lela Kogbara, one of GG's Trustees who was doing the MSc. "See you on Monday at the Trustees meeting," I said, and then as I walked away, I turned and asked, "Do you think it's a good idea to set up a circle of chairs in the big polytunnel at the Story Garden for the beginning of the meeting, at least while the light holds?"

"Yes, I think that's great," Lela replied, "but I am worried how the meeting will go with all of the tension and division that has been happening in recent months." In what seemed like a flash of insight, Lela suddenly said, "Rather than diving straight into logistics, perhaps we could do a 15-minute check in about how everyone is feeling, and what we hope for the meeting."

"Yes, that's a brilliant idea," I replied, "All I want for us is to create a space to really listen to each other and say the things that need to be said."

After we parted, I walked over to the Story Garden and set up the space in preparation for the meeting – it felt good to be there. I walked to the Oaks and stood in the middle of the circle, in the heart of the oak forest, curious what the trees might have to say to me. As I stood, I saw the wasps and the other insects and I felt the atmosphere. Trust was the word that came to me. I was buoyed up by the thought that unexpected breakthroughs can happen when a field of trust is created. I thought of the words of my colleague, Gill Coleman, who had been running the MSc workshops with me; "The only point of learning is if you can enact what you know." Now was the perfect time to practice this. When I got home, I began journaling and used a framework we had introduced to our MSc students to structure my reflections[iv], which helped me prepare myself emotionally for the meeting and write about it afterwards.

1. Intention

   Rather than meet what I perceive as patriarchal power with the same bullish force of patriarchy that also exists in myself, I want to approach the meeting as if it were a prayer, in that I want to be available for the sacred. I hope that before going into any negatives we can make space to appreciate each other.

2. Choices

   I knew I would have to put effort into letting go of control in order to give Lela the platform to take a lead in the check in.

   I chose to fill Nicole in on what had been happening.

   I thought about my appointment at the hospital beforehand. I decided not to bring this or any of the results from the appointment into the meeting. I also decided that if my appointment was delayed, I would leave the hospital; I needed to be on time.

3. Actual Action

   I was on time. In the circle in the polytunnel, in the twilight, amongst the last of the season's tomatoes a level of openness and intimacy was created.

   Lela opened the circle by asking, "Are you all up for doing a check-in?"

   "What's that?" Tony said.

   "It's an opportunity to relax and to get things on the table that might otherwise not be said," Lela replied.

   "That sounds good," said John, "Let's do it."

   Lela began and we moved around the circle. Rasila came next and then suddenly from across the circle, Tony said that it was important, in the interest of openness, for the other trustees to know that he would be stepping down at the end of the meeting. He explained that on returning to the trustee board he had not found it easy and did not feel able to continue.

Others spoke and there was a tangible sense of vulnerability and honesty around the circle. As I remember it, we shared appreciation as well as understanding and concern for the anxiety and misunderstandings that had been created between us. Amongst other things, it was acknowledged that my cancer diagnosis the previous year had added to the sense of uncertainty and the perceived need to lock things down. From then on, the meeting was straightforward and simple. John said he gave his full support to Steve being chair and that as of this meeting he was stepping down.

4. Feedback

   In the days after the meeting, it felt like a huge weight had been lifted off my shoulders, and I felt inspired to circulate the Trustees recruitment pack. Within days we had positive responses from two great candidates, one of them recommended by John. We also appointed Amir Miah to the board. Amir, who had stayed in touch with us over many years, had been a former Generator. Along with helping to set up a homeless charity, Amir was now a Masters graduate, working as a financial fund manager.

Nicole had been a constant companion throughout all of the ups and downs and critical turning points of the previous year. During that period, I had often said to my husband Rod, I am so glad I am not doing this job on my own, it makes all the difference that Nicole and I are on the same page. After reading several draft chapters of the book, including this one, Nicole wrote:

> I particularly liked reading and reliving the last year. As I thought about the tensions that came about with the trustees, it was good to know that we are now in such a different space again. It just shows that life is made up of bumps all along and that we get over the bumps and soon enough there will be new bumps again! … I also enjoyed reading previous chapters and seeing how our conversations have changed over the years as my position in the organisation has changed and as we have had more people coming in to join us. I feel that the honesty between us and open communication about the more challenging bits is what has made our joint directorship possible and successful. On a few occasions, I have had to share how I have felt undermined by your style of leadership and it is through having been able to have those conversations that I feel we have resolved this and found ways to work in which we are supportive and respectful of each other and our different ways of doing things. As a result of having been able to speak about this, I know that you have taken extra care to ensure that I can step into my own leadership. It has been important both for myself as well as how I am perceived by others. In the early days, the trustees looked to you, now that's changed.
>
> April 2020.

Now that we weren't being driven into a corner, Nicole and I felt free to seek the support we needed; support which the trustees had been encouraging us to get all along. Without much effort, we came up with a role description, advertised and before too long our new Assistant Director, Judy Hallgarten, had been appointed.

## Help on the ground

I picked up the phone and it was TJ, an active local resident who lived on the housing estate next to the Story Garden. When I first met TJ, she wore a number of hats; creative writer, committee member for Big Local community group, and Community Investment manager for Costain Skansa, a construction joint venture who were doing the enabling works for the new HS2 highspeed trainline. TJ describes herself as a parent, a disabled BAME woman, and a member of the Somers Town Community. "Can you come and meet my manager?" asked TJ, "Our company is wanting to put in a garden at the National Temperance Hospital." Thoughts of the dark Victorian building that was a feature of Euston's Hampstead Road came to mind. I wondered what the grounds were like, excited at the idea that there might be scope for a community plant nursery. A few weeks later, I made my way to the meeting and was disappointed to find no trace of the hospital, just an army of multi-story portacabins in a landscape that looked like a war zone. Numerous housing blocks and other buildings, including the hospital, had been demolished as part of the early works for the HS2 scheme. After the lengthy security process that is a feature of construction sites, we made our way up to a small windowless meeting room. TJ's direct boss and another senior community engagement manager began outlining their plans for what they described as a community garden. They wanted to involve Global Generation and although the signs weren't good, I tried to keep an open mind to what might be possible.

"Can we see the area where you would like to make the garden?" I suggested. We peered out the window to a small area bounded by the busy road and portacabins. "When would you like to begin? Have you been speaking to the people who live around here?" I asked. The answers came thick and fast, "It needs to happen in two weeks, and it will last for six months and no we haven't yet involved the community in the plans." The communication around the table seemed quite evasive; what involvement they actually wanted Global Generation to have was unclear as it seemed their subcontractors were lined up to do most of the work. Did they want our design and gardening input or perhaps it was an endorsement to call it a Skip Garden that they were seeking?

I took a few breaths, weighing up how best to respond; I felt wary and I wondered if I could say what I thought without killing the best of their fire? "I am afraid we couldn't put our name to it," I said, "It would be reputational suicide. Our whole approach is based on involving the community in making a garden, it is a way of growing together. We would like to work with you on a longer, slower project done another way, perhaps when you have finished with this garden, we could use the skips in our new Story Garden, and we could collaborate there."

TJ walked me out into the street after the meeting, "Thank you," she said, "I have been trying to tell them that we needed to involve the community from the get-go but no one listened." Months later TJ shared with me her frustrations about the meeting and why she had taken on the role in the first place. As someone who sought to contribute locally by standing on both banks of the river, I particularly valued her insight.

> I had been joining community groups for years who were already geared up to thinking the way that I thought and I was ready for a challenge. I felt that the best way for me to impact what was happening to my community was for me to have a say and be actively involved in the process. I wanted to bridge the gap, to present as someone that both sides could trust and communicate with in order to create effective community projects. But it wasn't easy, everyone was suspicious of each other and of my different relationships. The meeting for me was very uncomfortable because I could hear that suspicion, see the mistrust, feel the misunderstandings, from both sides. I was unable to strip away the suspicious silence and the cautious comments and help each side understand each other better. The only community engagement CSJV had been having with community groups up to that point was hostile which made them prepare only for contention and suspicion. I had hoped that we could tell them how to deliver and run this project with the community and the value of paying a grassroots organisation to do the initial work. They had subcontractors on board to create the garden because that's how constructions companies work, that's what they know, that's what's easy for them. As much as CSJV came to meetings prepared to be disliked, community groups came to discussions prepared to be ignored, belittled, taken advantage of. The difficult conversations grew from that foundation.
>
> TJ, May 2020

The garden on the National Temperance Hospital site did happen as planned. The ground was levelled and far smoother than anything Global Generation had ever achieved, and cyclamen, miniature conifers and other off-the-shelf plants were lined up like soldiers inside each of the skips. A sign over the arched entrance way said "Community Garden." Maybe people visited and appreciated the gesture, I am not sure. What I do know is that, true to their word, a year later ten empty skips were delivered to the Story Garden and this was just the beginning of an enormous amount of help and heart from the Costain Skansa supply chain over the next 12 months. Nicole and Martina were now the point of contact with TJ and others in her team who TJ had brought on board and I gathered from them that things were happening, but the scale of help went far beyond what I or any of us imagined possible. Each Wednesday morning, all of our staff come together for breakfast and a meeting where we share highlights and updates. On one particularly sunny Wednesday, as we sat upstairs in the Sky Room, our own portacabin meeting space. Nicole and Martina described how the week before they had sat in a circle in the Story Garden with about 20 people from Skansa's supply chain.

> Nicole: We were in the oak circle with a fire burning in the middle and we asked them why green spaces like this were important in the city. I thought they would each say a line but they all really opened up about their childhood, and all the things they are concerned about in the world, their own children and their children's children. It was really, really touching and very good practical outcomes are coming out of it as well.
>
> Martina: The Skansa day was completely different than how I expected it; I thought it would be quite corporate and they would all be in suits and there we were sat around a fire together. Before we had even started, they had already pledged support, to the point where they were almost fighting with each other over who could provide what. It is all moving really fast, on Monday I had three meetings from different people in the supply chain, we have got foundations for the classroom starting this week, they are drawing up all our site plans, we've been offered a storage shed. I have been inundated with emails from them asking how they can help. They are asking for nothing back in return, no one has mentioned logos which is really nice.

Skansa also shared their take on what proved to be a landmark morning for the Story Garden and in our relationship with the construction industry:

> Skansa: What a fantastic icebreaker! It was a great way to have people lower the "only open for business" façade. By sharing personal experiences, we got real insight into each member of the group and established a common interest in wanting to give back.

Months later, Martina and I stood in the pouring rain peering through the heras fencing at our nearly completed wooden roundhouse; beautiful and somewhat unbelievable that it had happened at all. Martina commented that this was the kind of project that would never be signed off, if one knew in advance all the steps it would take to complete. What began as a two-week participatory project, involving a large group of architectural students from the University of the Arts, was completed over five more weeks by Lowery's, one of the construction contractors who extended their help across the whole garden. I thought about the toilets, the electrics, the plumbing, and the outdoor kitchen. These were areas in the Story Garden that had been completed, and in some cases, rescued by the Skansa supply chain – Lowery's, Woodlands, McCain, Atkins, and Arup along with additional contributions by other unrelated construction companies like Gouldris and Waites. We had many things to be grateful for. As Ben Lichtenstein[v] writes, what actually sparks transformation is somehow beyond theory, unreachable through logic and not tied to rationality. The garden was now the expression of a huge joint venture; in many ways testimony to power of what he calls "grace, magic, and miracles." However, as TJ pointed out, the big difference on this project was that by the time the construction companies got involved the Story Garden already had hands-on involvement and support of the local community where as the National Temperance garden did not. The Temperance garden was always going to be a more difficult proposition given local sensitivity surrounding the HS2 project. Below are photographs of some of the Story Garden structures which would not have been completed without the support of our friends in the construction industry.

Our decision to work with the coalition of companies that are involved in the enabling works of HS2 could well be viewed as questionable. The installation of this high-speed train line is having a big environmental and social fallout and is

Community roundhouse in the story garden, finished the week before lockdown – photographer Sarah Ainslee

Inside the roundhouse, recording with theatre complicite for our voices of the earth project – photographer Sarah Ainslee

Story garden office and community kitchen made of two shipping containers with the Francis Crick Institute in the background – photographer Kiloran Ben O'leary

not universally popular; some would go so far as calling the companies involved "ecociders." The way I see it, HS2 is happening whether we like it or not, like TJ, I feel that by working with, or at least having dialogue with the companies involved we can be part of helping them restore some of the damage done, in this case in the area surrounding Euston Station. I am committed to a way of working that does not draw strict lines between "us" and "them"; as described in previous chapters, Global Generation has collaborated with socially and environmentally questionable industries and I have been open about what I think of some of their projects. On more than one occasion, I have witnessed how growing trusting relationships can help to seed fundamental shifts in the kind of worldviews that treat the planet as if it were not alive and a fundamental part of who we all are.

## Reflections and final comments

In the far corner of the Story Garden is a huge, bright pink agricultural bowser that accompanied the Beuys Acorns into the garden. Heather and Dan had bestowed the bowser with its new coat of pink as a tribute to the pink boat and thousands of rebels who had gathered to demand the declaration of a climate emergency in the Easter of 2019. When I see the bowser, I ask myself, what is our contribution in the story of a climate changed world? The question is open and ongoing as it should be; how can any of us pretend to really know? The little I do know is that the Story Garden is providing sanctuary; for people and for the more than human world. As I wrote this, the oak buds were beginning to prick outwards, the bees were returning and the foxes becoming tamer; their coats once mangy, now thick and healthy. Perhaps they were all the better for the circles of shells children had placed at the entrance to the burrows they had made in the piles of soil and woodchip.

Doing very public work within the constraints of the institutional world, we needed to be able to speak the language of that world. In so many ways, the success of a project comes down to communication. Only by speaking the language of the system could we challenge its preconceptions about what might be possible. Identifying and discussing common ground when you work together is a really important starting point and for that you need to know each other's language. We needed to gain trust to unlock privately owned spaces and to secure planning consent along with raising funds. At the same time, we needed to earn the trust of the local people who lived around the garden. For Global Generation, this was made possible through the dedicated input and hard-earned experience of local champions like TJ and many others, along with the creativity and understanding of a flexible and forward-thinking architect like Jan and our own, very proactive, in-house community build manager, Martina, who has an architectural training. Less visible and absolutely crucial, legal support was again offered pro-bono by Ben Jones from the law firm Freeths (and previously Herbert Smith). Ben, a property lawyer, has skillfully supported Global

Generation's projects for the last ten years or more. He holds his craft lightly, and in doing so has taught me to appreciate the power of the legal framework as a background layer in a path-making process. As with the development of the Skip Garden (written about in chapter 2), we married legal and professional requirements with community-based understanding and experimentation. On both projects, time constraints and institutional realities meant that the ideal scenario of occupying a site and listening to how the land and the local community might speak to us over time, throughout the changing of the seasons, was not an option. Consequently, the level of communication these projects really deserve was compromised. This is how Martina described her experience after reading a draft of this chapter:

> I'm very grateful to Jane and Nicole for sticking to their instincts and trusting in a more flexible process of development which from the outside could be seen as a risky strategy. I have thought more about the tensions caused by the decision not to have the main "road" in the Story Garden. If this conflict of ideas happens again, I think we will be more prepared to make the time to enable multiple voices to be heard and to discuss each idea fully. This is what I understand to be at the core of a successful "commonplan" model, which is a new way of working to some of us. Although maybe bumpily at times, I feel we are navigating our way through to find a way that works and is truly collaborative. This reflexive learning curve is in the spirit of all that Global Generation encourages as an organisation; from listening to others as well as our own inner voices, taking time to be still and understand, letting go of ownership or ego to allow space for the new.
>
> Martina Mina, May 2020.

A project like the Story Garden raises many questions; what range of skills are needed? What is an appropriate balance of pro-bono, volunteer, and paid input? How best to inter-mingle and transition between professional input and the more emergent, on the ground community-based processes? Does leadership of different aspects of the project need to be singular or collective or a combination of the two? How much information to share and when. On reading this chapter, Kiloran, the Story Garden manager, said, "I am very glad I didn't know what was happening behind the scenes with the trustees, I think I would have lost confidence if I knew they were worried about what we were doing." I would say it is very difficult, if not impossible, to know the answers to all these questions in advance. My experience is that many possibilities in a "commonplan" process occur through happenstance. For example, when we abandoned the straight lines of the brick path, we had no idea that a few months later the Beuys Oaks would take its place, bringing a magical presence and a welcoming space for so many gatherings like the one in the photograph on next page.

Storytelling in the Beuys Oaks with Dontae Jacobs and Molly Frow and Naomi Frederick of theatre Complicité

Along with allowing time for proper communication, what one can do is cultivate the ground by setting:

- a foreground programme of hands-on community activities in which the power of relationship can grow.
- a background framework of legal and social agreements (if the site requires it)
- a basic infrastructure for operation e.g. plumbing and electrics, toilets, kitchen and an office.
- a culture of storytelling, silence and other ways of invoking the mythological heart of a forest.

Through projects like the Story Garden we are not just contributing according to the more tangible physicality of our charitable aims but, as one of our Trustees, Jane Jones, described, we are helping to pioneer a different way of working, that focusses on "I" and "We" as well as the "Planet." We are doing this not only for ourselves but in the hope that we might lay down possibilities for others in the future.

> In a climate changed world where the traditional structures and ways of doing things will increasingly, and probably quite quickly, be forced to shift as we try to adapt to the many disruptions heading our way, GG is one of the few organisations I am aware of which has been brave and forward-thinking enough to start to live by a new paradigm and

thus show the way to others. This can only have been possible with the backing of an equally brave and forward-thinking board. It's one of the reasons I was attracted to become involved in the first place, and while compliance with legal and governance requirements etc. is evidently crucial (and not to my understanding a problematic issue for GG), I think this "brave new world" paradigm is just as important for the ethos of the charity. For me, values and ways of working are inextricably linked and I think it is therefore important that the board and the executive continue to be aligned in this respect in order to work most effectively together.

Jane Jones, October 2019.

Diversity of views on a board is important, and finding one's way through conversations where worldviews and ways of working collide, anxious-making. I found it a delicate balance of listening and not being too rigid in my beliefs, whilst not compromising on the desire to align all parts of the organisation to the underpinning "I, We, and the Planet" philosophy behind the work. I didn't always get it right, experienced many sleepless nights worrying and wrote at least one apologetic email because of my behaviour. The tendency is to remember the break-through moments and yet the real work is everything else that lays in between. Greg Boyle, who does mighty work in the heartland of America's gang culture in Los Angeles, describes the predicament of promoting false reality through our narrowly defined victory stories. Quoting Mother Teresa, he says "we are not called to be successful but faithful."[vi] For me, being faithful is about sticking with the work through thick and thin and being willing to learn from the bumps as well as the breakthroughs. Whilst something of a cliché, in my experience the difficult times were the ones that delivered the kind of questions we needed in order to deepen and grow. The fact that Nicole and I and other colleagues were "faithful" in our approach eventually brought the resources needed to build the garden. Lichtenstein[vii] describes how transformation occurs out of shared vulnerability and must emerge out of the elements and personalities that exist in the group. The vulnerability of the trustees in the polytunnel meeting and the way it came about left a lasting impression on me, and I hope gave confidence to everyone, including those who had struggled to identify with the more intangible strengths of an organisation committed to standing in and evolving with uncertainty. Over the first year of the Story Garden, three board members did step down, but they did this with grace and in good spirit, with a legacy of good work behind them. We learned to be more upfront about the importance of leading together. This was clearly stated in the role description that attracted three new Trustees who replaced the ones that left: "GG is pioneering a way-finding, action research approach to its development, championing collaborative ways of working within the team and moving away from traditional, hierarchical organisational models. New trustees should be open to working within this paradigm."

In addition to creating change within the board, it was important to keep working on the internal dynamics amongst the staff team. Jane Jones ran a series of powerful workshops for our staff and associate team. Drawing on her professional experience, both as a Human Givens psychotherapist and as a Programme Director with leadership development organisation "Olivier Mythodrama," Jane introduced the team to the dual concepts of human emotional needs and archetypes. [viii]Through a series of creative activities, we were able to self-assess our emotional needs at work and identify our preferred archetypes (for example; nurturer, warrior, explorer). This gave us a shared language to understand and relate to each other in new ways, and to identify and talk about our differences in a less personal way, particularly those that caused tension between us. In my case, I self-identified as a "sovereign-renegade" (strongly holding a vision for the organisation and wanting to do things in new ways), and recognised that my emotional need for "safety" at work was met through creative freedom and innovation. I had on more than one occasion clashed with Gwen Mainwaring, our head of events, who had responsibility for health and safety in public gatherings at the Story Garden; Gwen self-identified as a strategist (an archetype with an eye for detail and a concern with order and procedure), and her need for "safety" was met through making sure things were being done according to the rules. Sometimes I would say "yes" and Gwen would say "no" in response to clients' requests, which had caused tension between us. The work with Jane helped us see that we were viewing such situations through different lenses because of our different archetypal preferences and the ways in which we each met our emotional need for safety. This realisation helped us both to understand each other's perspectives and dissipate the tension between us. Through the work with Jane, I thought more about how I can be undermining and unhelpful, in that that if someone else is posing what I consider to be obstacles to my intentions, my initial impulse is to think, I'll just get on and do it myself. The renegade, or in the language I have used in these pages, the "pioneer" in me needed to slow down, listen, appreciate the emotional needs of others, and embrace a more collaborative way of being (all characteristics of the nurturer archetype). Through the understanding of archetypes and emotional needs, I became curious rather than insistent and was able to appreciate the difficulties Gwen was up against, particularly when it came to planning our first large event in a new, unfinished, and not fully operational site.

Writing about these events has deepened my appreciation for the power of action research to bring about change. It has sometimes been a painstaking process moving from me to we; naming people, writing about tensions, sharing half-baked chapter drafts and then inviting those I had written about into the writing; often long after the event. Weaving in other people's perspectives has developed my understanding and loosened the negative hold of past events upon the present. Whilst there is always further to go, it is the diversity[ix] between us that creates an organisation that I compare to a woven wicker basket; open, flexible and strong.

## Postscript

Glancing skywards I receive a thumbs up from the window of the passing 390 bus. "That's great," "Thank you," "What a good idea," were the responses of the people who passed by. It was somewhat surreal to feel so uplifted in this extreme circumstance. This was no ordinary Saturday; we were just a few days into more intensified measures set out by the UK government in response to COVID-19. Whilst we weren't in complete lockdown, all of us were doing our best to keep two metres apart. It was like one of those awareness building activities, when you are asked to move around a room without actually touching. People who stopped to talk to us stood a good distance away. We were clambering in and out of the skips pictured below, which were creating two 'parklets' in the car parking spaces on the side of the road. We were drilling, sawing, moving soil, planting and taking care not to spread the virus. I was with my neighbours who are all part of Greening Leighton, a campaign to make Leighton Road and the Kentish Town streets surrounding it a greener place to live.

On this particular day, we were filling with soil the big space in the middle of the skips. The central opening is surrounded by shallow wooden growing beds that were now blooming with tulips and daffodils. Having been involved with the mobile King's Cross Skip Gardens for the last ten years, initially I wasn't so keen on this new design. "How long will the Skips be there for? How will we

Skip Garden claiming the street in Leighton Road Parklet May 2020, Kentish Town – photographer Jane Riddiford

move them with all that soil inside?" I asked. A few words conveyed the vision and commitment behind what was happening; "We want them there for ten years or more … maybe forever," came the response. What an ideal destination and legacy for the Skip Garden, I thought. This was community-based regeneration at its best; bottom up and spontaneous. What's more, it was being done for next to nothing; nearly everything was found or donated, people were sharing tools and bringing cuttings from their gardens.

Already a growing arch out of found piping had been installed in one of the skips. A wire had been cunningly stretched across the pavement to a nearby garden, this would become a perfumed Jasmine walkway. During the three hours that we worked together, there was talk of how we might green the remainder of the street-side parking spaces that had been allocated to our efforts. A pallet garden, a community composter, water butts from the down pipes of the nearby houses, were just some of the ideas.

During the afternoon I picked up several emails from my colleagues in Global Generation. A few miles down the road in King's Cross more collaboration was happening. Local organisations were reaching out to each other and evolving a shared response to the challenges that COVID-19 was presenting us with; more food growing, safe possibilities for nature connection and respite in the Story Garden, making sprouting kits for local families, online possibilities for the children and young people to learn and share their experience of these changing times. At the time I was reading Rebecca Solnit's book; A Paradise Built in Hell[x], which tells the stories of some of the small localised utopias that occurred in the midst of major disasters; from the 1906 San Francisco earthquake and fire to Hurricane Katrina in New Orleans. Solnit describes how embedded in the word emergency is the emergence of new possibilities; new ways of being and living together. Possibilities that even a week before I would have found hard to imagine.

> *From my Journal – from another time.*
>   *Looking at a leaf, I glimpse a different world*
>   *I am not who I think I am*
>   *I breathe not what I think I breathe*
>   *The land under my feet is old*
>   *I know not what I write*
>   *The story is not made by me*
>   *Drop the flapping skin of the hag*
>   *The crude and beguiling mask of vanity*
>   *Write no more until there are*
>   *Words to sing of sorrows on my brow*
>   *the un-tempered quest for what lies within*
>   *Please nobody*
>   *Do what it is you are to do*
>   *Stand still, the forest will not forget you*
>   *Listen to the Moorpork calling*

## Notes

i     Snyder, G. (1990). *The Practice of the Wild*. Berkely: Counterpoint.

ii    Harrison, R. P. (1992). *Forests: The Shadow of Civilisation*. London: University Press

iii   Odonohue, J. (2012). *The Four Elements: Reflections on Nature*. Ireland, Transword.

iv    Action research encourages the study of different territories of experience: inner, outer and collective. An action researcher also considers the different time frames in which our experience is situated; past, present and future and how these intersect. Bill Torbert offers four headings as prompts to stimulate this way of inquiring; Visioning, Strategising, Performing and Assesing. Words, including Torberts, change through use. We introduced these headings to our MSc group as: Intention, Choices, Actual Action, Feedback. See Torbert, W. (2001). The Practice of Action Inquiry. In P. Reason & H. Bradbury (Eds) 2001. *Handbook of Action Research* Sage: London. pp. 250-260.

v     Lichtenstein, B. (1997) *Grace, Magic and Miracles: A chaotic logic of organisational transformation*. Journal of Organisational Change Management. Vol 10. pp. 393-411

vi    Boyle, G. (2010). *Tattoos On the Heart*. New York: Free Press p.167

vii   Lichtenstein, B. (1997) Ibid.

viii  For more information see https://www.oliviermythodrama.com/

ix    It is a well-known fact that many environmental organisations are predominantly white and middle class. As I described in chapter 2, Global Generation was co-founded by Paul Aiken, a Jamaican, and supporting diversity in both land and people was at the heart of our founding mission. However, as our staff team grew, sadly we became less diverse. During the period the Story Garden was developed we applied ourselves to growing a staff and board team that would look like London, reflecting the diversity of the children, young people and families who are involved with us.

x     Solnit, R. (2009). *Paradise Built in Hell: The Extraordinary Communities that Arise in Disaster*. New York: Penguin.

# APPENDIX 1: OPPORTUNITIES FOR YOUR OWN INQUIRY

## Chapter 1 – An inner-city forest

If you have got this far, your journey in the lattice of stories in this book, and hopefully in the stories of your own making, has begun. In the spirit of beginnings, I invite you to honour the land and the stories of your own ancestors. Take some time, either with a friend or by yourself to recall the lineage that has shaped you. Start with a minute or two of silence. Begin where you naturally land, long ago or the recent past, don't think about it too much. Bring to mind your ancestors, be they the stars or your grandparents. What was the land your family came from, what were the journeys they made? Where are the special places in nature that you spend time? You may choose to do this through writing or by gathering symbolic objects. Find an opportunity to share and to take part in the practice of mihimihi with others.

## Chapter 2 – From rooftops to developers' land

It's easy to feel that we are supposed to exist in some kind of sparkling eternal summer which we can control. As any gardener knows, there is potency in the resting and restoring rhythms of the seasons and in the unpredictability of exactly how and when these cycles will enact themselves. The original meaning of the word crisis is change. Children's author, Jennifer Morgan[i], describes how in times of crises the universe gets creative. In the depths of the ocean, it took two billion years of trying and no doubt missing before cells managed to find a way to come together to form complex groups of cells out of which the diversity of life has come. Identifying with the patterns in nature as our starting point has enabled Global Generation's gardens to consciously grow within uncertainty about where we will end up.

Key to the ability to embrace what can at times be an anxious-making process has been taking time to reflect and habituate ourselves to ways of being that run counter to the dominant narrative of certainty and control. One of the ways we have learnt to inhabit this bumpy ground has been through engaging in "Freefall" writing, first introduced to me by Barbara Turner Vassalego[ii]. It is a practice that is now a regular feature in Global Generation's workshops with children and young people. The writing they produce bypasses limitations of the editing mind, eliciting a more spontaneous reality onto the page. This is an invitation for you, the reader, to do some freefall writing. Use a favourite notebook or a decent piece of paper and find a good place to sit. Take a few minutes to be still and silent and then when you are ready, begin writing about a time in your life where things didn't exactly go according to plan. What creative actions emerged as a result? You might use the start line: "The sun was shining and then ... 'Write freely without reading back on what you are writing. Follow the energy of the curiosities or the resistances that arise in your mind and bring your whole experiencing self to the process. Invite the richness of all of your senses to your words. What did you feel, hear, see, or smell in this time when things didn't go exactly according to plan?'"

## Chapter 3 – Leading as a way of being

I invite you to find an object in the garden or at home that in some way speaks to you of the values you hold dear in the context of leading. Once you have found your object, use a freefall writing approach to reflect on a situation that helped you deepen, develop, or even abandon those values. Remember to include all of your experience; your heart beating, the feeling in your gut, the gravity in your bones. What was the scene, the colours and the textures, the sound of the voices. Giving all of the sensory detail will take the writing deeper and unlock unexpected insights about your experience.

## Chapter 4 – I, we, and the planet

The birth of Global Generation came from the experience of taking young people on camp at Pertwood Organic Farm in Wiltshire which continued over many years. Early morning walks across the Wiltshire downs have been a powerful way of evoking the depth of silence. Before breakfast, we walk in silence, with occasional spoken guidance from a facilitator; "Feel the ground under your feet, the wind on your face, the whisperings in the air." There is a sense of when the group should walk together in formation or when they should spread out across the fields finding their own ground, each one getting a sense of themselves without being busy with the others. I often speak about building internal muscles, an inner source of strength that is independent of what we think and even what we feel. When we come together again, usually in a large circle, the silence between us is loud and palpable. In this atmosphere, the question of responsibility emerges naturally; it comes from the inside rather than being externally imposed.

> *As we are all part of the whole, the question it raises for me is – what is my role in it?*
> Moses Kasibayo – 21 years old, after a silent walk

I invite you to get up early and go for a silent walk, ideally in a park or green space or perhaps by the sea. Alternatively, you might discover that, when done with intention, even the harsh terrain of the cars and the concrete can take on new and magical dimensions when met in silence. You might choose to do this on your own or with a friend. Walking in silence with others can be a powerful way of entering the silence. It might help to choose how long you are going to walk for. I recommend at least 20 minutes to allow yourself time to settle, but it could be for up to an hour or more. As you walk, ask yourself the question; What does the land have to say to me? When you return, take time to do some freefall writing in your journal with the start line – What did the land say to me?

## Chapter 5 – A cosmic story

Take some time to think about the stories you heard as a child about the origins of the universe. Do some journaling or discuss with a friend how these stories made you feel. What do you think of them now?

The story of the creation of the universe is an enormous story and if it is held in a fluid and open way, it is big enough to include many smaller stories. In this regard, it is very much a community-building story that can provide a unifying thread between different cultures and also between the past, the present, and the future. I invite you to find your way into this story. From the timeline below, choose a part of your deep time history that speaks to you. This is by no means a conclusive list; you may choose to create your own deep time history. The main thing is to have a sense of a whole journey from our distant origins to our current life time. You might even want to include your hopes for seven generations into the future.

> *The original flaring forth - 14 billion years ago*
> *Birth of stars and galaxies - 13 billion years ago*
> *Milky Way galaxy - 12 billion years ago*
> *Super Nova - 7 billion years ago*
> *The sun - 6 billion years ago*
> *The earth - 5 billion years ago*
> *Bacterial life is formed in the bottom of the oceans - 4 billion years ago*
> *Eukaryote Cells, complex cells are formed - 2 billion years ago*
> *Early sea creatures - 500 million years ago*
> *Dinosaurs - 200 million years ago*
> *A giant meteorite wiped out the dinosaurs - 65 million years ago*

*Development and spread of mammals - 60 million years ago*
*Rainforests - 50 million years ago*
*Whales emerge in the sea from returning land mammals- 40 million years ago*
*Apes appear - 15 million years ago*
*Hominids - 7 to 8 million years ago*
*Agriculture - 10,000 years ago*

The present day

Spend a few minutes imagining the part of the story you have chosen and contemplate how it is actually part of your own history, as that part of the universe how did you feel? What did you imagine? What qualities did you have? Were you fiery or quiet and calm? What did you see, smell, taste, and touch? What happened? Writing in the first person, begin with the start line, "Once upon a time I was …"

Feel free to let your writing take you forward or back into other parts of the story. Relax and let yourself be surprised.

## Chapter 6 – Encounters with the high priests

Take some time to reflect on an unhelpful leadership dynamic you have been involved in, either as a follower or a leader. Notice what qualities were at play in yourself and in others. Identify a question about leadership or followership you carry going forward. You could try essentialising the forces at play by writing about your experience as a mythic story.

## Chapter 7 - Listening to land

Think about a formative event in your life, either as a child, a teenager, or an adult. Write about or describe to someone else what happened. Try to describe the scene. Where were you? How did you feel? What happened? When you have written your story, think about what enduring qualities or values it has shaped in you. Now share your story with someone else and ask them what qualities they hear in the story.

Take some time outdoors in a garden or a park. Walk slowly and find a place that attracts you. Cut your pace in half and listen for something that is calling you to stop and pay attention, it might be a tree or a leaf, the soil or the sky, or the feeling of the wind. Write an ode, a letter of appreciation to whatever it is that you are spending time with. When you have finished writing, you contemplate how your ode, resonated or not, with the story that has surfaced for you about a formative time in your life?

## Chapter 8 – Growing a Paper Garden

Throughout the workshop process, in which we developed the Paper Garden, colleagues and I shared traditional stories and mythical styled stories that we wrote ourselves which resonated with the themes we wanted to explore with children and young people. I invite you to explore creation or other traditional myths from your own cultural background. Is there one that resonates with you and that relates to the locality where you live and work? Learn the story and share it with a friend or a group of friends. In this way, you will be enacting an age-old ritual of community-building.

## Chapter 9 – In the jaws of the corporate dragon

The stories in this chapter describe a process of holding a vision and letting go of fixed ideas. As others have described, the empty space is the creative space where imagination and new possibility can really grow. Rilke invites us to "Know the great void where all things begin."[iii] Matthew Fox claims that an important aspect of any work is to develop a practice of emptying, of embracing nothing-ness out of which authentic work in the world will arise; work that he claims is both a source of enchantment and is aligned to the codes of the cosmos.[iv] Within that empty space, magical stories can grow which offer vision and strength to create new realities.

Take a few minutes to be still and silent. Think of a time in your life when things seemed impossible and you found a way through by letting go of your pre-conceived ideas. What did that feel like? What did you do? What happened?

Think of something you would like to do in the future. In your journal or to someone else, tell the story in mythic proportions; imagine it is possible, give your characters all of the colour and magical powers of the heroes and heroines of fables and legends.

## Chapter 10 – From the heart of an oak forest

Tiny moments of connection inside the heart of the "oak forest" in the Story Garden had an impact on not only me but those in the communities we work with; the Bangladeshi Mothers, the construction employees, my colleagues in Global Generation. In their own ways, the oaks and the many plants surrounding them spoke to each of us, with messages of connection, trust, and care. In all of these moments, it was not what we knew about the plants but rather their ability to help us shift gear into a deeper kind of listening. A way of being in which all of the natural world, and perhaps even the rocks and stones themselves, might be experienced as alive; a way of being in which to hear and honour the voices of the earth.

During the first year of the Story Garden, colleagues began a project called Voices of the Earth, in collaboration with a number of local partners: Theatre Complicite, The British Library, The Royal College of Physicians and Hopscotch, an Asian women's network. In this exploration into the healing power of plants, through history, culture, and our own experience, very soon it became apparent that we were involved in democratising heritage. By that, I mean privileging our own experiential ways of knowing alongside institutional knowledge. In this contemplation, I noticed that, more than their physical attributes, it was qualities that the plants seemed to evoke for each of us which were most transformative. One could say, the inner dimension of plants met the inner dimension of people.

If you would like to honour the voices of the earth, I invite you to walk out to place you feel a connection with; your garden, a park, a leafy street. As you walk, try and let go of other thoughts and be open to the plants you are drawn to; the dandelion, the dock, or the tree that you have no name for. Step towards whatever plant you are attracted to and spend some time there. Step away with gratitude. You may want to take a photo, or pick a small piece of the plant. Take the time to draw its shapes and lines, and as you do so, pay attention to the way that it moves you, and write about your experience.[v]

## Notes

i   Morgan, J. (2003). *From Lava to Life: The Universe Tells our Earth Story.* Nevada City: Dawn Publications. Set within a context of awe and wonder as a doorway into environmental responsibility, this is one of three children's books written from the point of view of the universe.

ii  Turner Vassalego, B. (2013). *Writing without a Parachute.* Bristol: Vala.

iii Rilke, R.M. (1989). *The Selected Poetry of Rainer Maria Rilke.* (Trans.) S. Mitchell. New York, NY: Random House, p.136.

iv  Fox, M. (1994). *The Reinvention of Work: A New Vision of Livelihood for Our Time.* San Francisco, CA: Harper Collins.

v   These guidelines were influenced by the instructions on how to meet a plant from the Intuitive School of Herbalism, passed on by Charlotte Gordon, Global Generation Community Gardener see https://www.schoolofintuitiveherbalism.weedsintheheart. org.uk/

# INDEX

Page numbers in *italics* refer to figures and those followed by "n" indicate notes.